The Physics of Cosmic x-ray, γ-ray, and Particle Sources

Topics in Astrophysics and Space Physics

Edited by A. G. W. Cameron, *Yeshiva University*, and
G. B. Field, *University of California at Berkeley*

K. GREISEN   The Physics Cosmic of x-ray, γ-ray, and Particle Sources

The Physics of Cosmic x-ray, γ-ray, and Particle Sources

KENNETH GREISEN

Cornell University

GORDON AND BREACH SCIENCE PUBLISHERS

NEW YORK   LONDON   PARIS

This book originally appeared as a section of *Astrophysics and General Relativity, 1968 Brandeis University Summer Institute in Theoretical Physics*, Volume 2, published by Gordon and Breach in 1971. The work was supported in part by the Air Force Office of Scientific Research under Grant No. AF–AFOSR–255–67.

Preface

Astronomy has undergone a tremendous expansion in variety of the sensing devices employed, in the spectral range of its information channels, and in the breadth of science required for synthesis of the accumulating knowledge. Before 1950, the field was essentially limited to what could be deduced from radiation in the narrow, visible band of the spectrum, emitted from stellar surfaces and clouds of gas that are warm by human standards, but not much hotter than a welder's arc, where the particles have typical energies of about one electron-volt. In 1970 the width of the investigated spectrum was no longer a mere factor of 2, but a factor exceeding 10^{20}, and included radio waves, infrared and ultra-violet, x and gamma rays and cosmic-ray particles. Most of the added channels of information are a direct consequence of the behavior of particles with energy up to a million million electron volts. Thus, high energy physics has become a vital branch of knowledge of the astronomer. Furthermore, many physicists have been drawn to astronomy as a more diverse and unrestricted high-energy laboratory than any which man can construct.

The rapidity of change in these fields has created a difficulty in keeping abreast of the state of the art, particularly since the new areas and techniques that have been embraced are themselves quite complex. An urgent need has therefore arisen for contemporary surveys: articles or small books that interpret and integrate knowledge from many sources, attempting to be broad in scope and viewpoint, but not excessively burdened with detail; to serve as an introduction for readers whose specialized prior knowledge includes only a small part, if any, of the subjects. This is the aim of the present volume.

The impetus for its composition was provided by the presentation of a series of lectures on the physics of cosmic x-ray, gamma-ray and particle sources at Brandeis University in the summer of 1968. The chapters of the book adhere roughly to the order and content of the lectures, but were written during 1969 and include new research findings that were reported in that year.

More recent results and interpretations of both old and new observations have appeared between the writing and the publication of this book, and will continue to arise, as is inevitable in so active and fascinating an area of research. A particular example is the recalibration of the OSO-III gamma ray detector sensitivity, which has led to a downwards revision by a factor of four in the high-energy gamma-ray flux reported in Chapter 6, and has substantially reduced the discrepancy between the data and the predictions based on known processes, as presented in that chapter. Impressive discoveries have also been made regarding x-ray sources (Chapter 7), stimulating a fresh enquiry into their basic mechanisms. As the years go on, the number and seriousness of such alterations in knowledge and outlook will grow until this volume has lost its usefulness, probably before 1980. Before then, we trust that its functions will have been taken over by a more worthy successor.

June 1971   KENNETH GREISEN

Contents

1.   INTRODUCTION: SOURCES OF HIGH-ENERGY PARTICLES

It is now 55 years since the discovery of cosmic rays, and 40 years since it was recognized that an overwhelming proportion of the primary energy is conveyed by charged particles. In the intervening years, much detailed information has been accumulated about the behavior and nature of these particles. The composition has become rather well known—that is, the relative abundances of the different nuclear species and of the electrons and positrons. Also the energy spectra of these species at the top of the atmosphere have been fairly well established, and extensive studies have been made of the variability of the particle flux with time.

Interactions of the particles with the media through which they propagate, however, have done much to obscure the sources of the particles, and even their composition and spectra at the sources.

1.1.   Alterations of Cosmic Rays between Source and Detector

The first barrier was the atmosphere of the Earth: the particles arriving at sea level bear little resemblance to those striking the top of the atmosphere. Most of the energy reaching sea level is carried by muons, an unstable type of particle not present in the primary radiation or in the material of which the Earth and stars are composed. The most prominent other particles reaching sea level are electrons, but not electrons of the primary radiation. Moreover, these muons and electrons are not even created by the primaries directly, but in two or more steps (beginning with pion production and decay), involving particles and mechanisms that were originally unknown.

The obstacle of the atmosphere has now been overcome in two ways. One was by acquisition of knowledge about high–energy physics:

1

the various kinds of secondary particles and the transformation processes and cross sections involved in their generation by the primaries. The other was by making observations above most of the atmosphere, first with balloons and ultimately with rockets and satellites.

A second barrier to simple measurement of the primary cosmic rays was the Earth's magnetic field, which distorts the directions of arriving particles and prevents many of them from reaching the top of the atmosphere at intermediate and low latitudes. The threshold momentum per unit charge for vertical arrival at geomagnetic latitude λ is approximately $15 \cos^4 \lambda$ GV/c. This threshold variation has been turned to advantage in the analysis of the primary spectrum. However, the field has other effects which have been troublesome as far as measuring the primary intensity is concerned. One is that the Earth's field gives rise to the radiation belts, in which satellite measurements of primary particles are made very difficult by the intense trapped flux. Secondly, the Earth's field is distorted by the solar wind, greatly complicating the analysis of arrival pathways open to the primary particles of low energy. Thirdly, interactions of cosmic rays in the atmosphere create an albedo (outgoing) flux of scattered and secondary particles, almost as numerous as the primary ones. The magnetic field deflects the albedo particles that are below the geomagnetic cutoff momenta, making these particles return to the Earth, usually at a similar latitude in the opposite hemisphere. The incoming particles, therefore, include a substantial component of re-entrant albedo. Many of the early measurements did not contain adequate corrections for this effect, and hence yielded reports of flux values that were not only too high, but also distorted in respect to spectrum and composition.

Above the Earth's atmosphere and magnetic field, a third barrier exists: a rarefied solar atmosphere, in the form of a gusty wind. The Sun is continually blowing its top at a highly variable rate. The exuded medium is a plasma sufficiently conducting to transport magnetic-field lines. The dominant outwards motion of the plasma draws out the field in a generally radial pattern from the Sun, but turbulence introduces irregularities and knots in this field structure. For particles of momentum under a few GeV/c, the Larmor radius is very small compared with the scale size of the system. The result is that such particles (the bulk of the cosmic rays) undergo anisotropic diffusion in the frame of reference of the moving gas, the outward convection of which transports solar particles toward the Earth and continually impedes the arrival of galactic cosmic rays. Thus modulation occurs: the flux arriving at the Earth is not simply related either to that generated at the Sun or that which exists in interstellar space. Both sporadic and secular variations of the solar wind greatly complicate the problems of observation and of disentanglement of the data.

At energies larger than 10 GeV per unit charge, the solar modulation effects are very small. Below this energy, a theory of the modulation effects has been developed, notably by E. Parker (1958, 1965), giving a prescription for their evaluation which seems reliable at least for nuclei of near-relativistic velocity. The existence of these effects not only creates an inconvenience in the analysis of primary cosmic rays, but also provides a tool with which to study solar activity and the interplanetary medium. Projected space missions to both inner and outer planets offer great promise, both of resolving the cosmic-ray problem and of providing rich information about the dynamics of the solar atmosphere. Meanwhile, however, quantitative statements about the low-energy part of the cosmic-ray spectrum and composition are rendered uncertain by the complexity of the modulation effects and the diffusive mixing of the solar and galactic flux.

Beyond interplanetary space, interstellar gas and magnetic fields cause further changes in the cosmic rays. The fields so confound directions of motion that the arrival directions give no observable indication of discrete sources. If the cosmic-ray density is much greater in the Galaxy than in intergalactic space, as is supposed, the majority of the particles we observe should be produced in the thin disk of the Galaxy, and there should be a generally outwards diffusion both along the arms and across the disk. However, the scattering length is so short compared with the dimensions of the system that the anisotropy is barely detectable (partly because of confusion with effects of the solar wind, and possible effects of galactic rotation as well as secular motions of the Sun and gas clouds within the Galaxy). Therefore in most cases the identity of the sources of cosmic-ray particles has to be inferred indirectly, by detecting secondary electromagnetic radiations (radio signals, or x or gamma rays) that emanate in straight lines from the sources and are produced there either by the high-energy particles themselves, or in the processes in which the latter are generated.

For instance, high-energy electrons interacting with magnetic fields in the sources can give rise, by magnetic bremsstrahlung ("synchrotron radiation"), to radio waves of a characteristic spectral shape and possibly strong polarization. Where the electrons have sufficiently high energy, these emissions can extend in frequency to the optical range and even to x and gamma rays. Fast electrons can also scatter the electromagnetic radiation in the sources and thus produce recoil photons of high energy. Heavy particles such as protons do not engage in radiative interactions appreciably, but interact with the ambient gas via the strong nuclear interaction, producing neutral pions that decay into gamma rays. These have a spectrum that peaks at half the pion rest-energy (i.e., at 68 MeV). The same type of interaction also makes charged pions that decay into muons and thence into positrons and electrons, which in turn engage in the afore-mentioned radiative processes. Thus, high-energy electrons and nuclei can make their presence known in remote places, even though propagation of these particles to the Earth would destroy the directional information by which the source might be identified.

Not only directional information is lost in the propagation of charged cosmic rays. Also the spectrum and composition are affected. Heavy nuclei undergo spallation and are reduced in number. In the same processes, some light nuclei are created that were not present in the sources. The loss of energy by ionization is proportional to Z^2 and is a strong inverse function of velocity, so the low-energy part of the spectrum is distorted in shape, and the distortion is different for different nuclear species. A low-energy electron component is produced by the knock-on process. High-energy electrons lose energy by synchrotron radiation and by scattering photons, at a rate proportional to the square of the electron energy, thus changing the shape of the spectrum. Furthermore, for all the charged particles, the rate of diffusion is probably energy-dependent, producing different spectra as well as different particle densities in different regions of the Galaxy.

These complications, like those mentioned earlier, can be turned to advantage for estimating the age of the cosmic rays and the nature of the diffusing medium. But they render the gathering of data and the determination of source characteristics more difficult.

1.2.   Guaranteed Sources of Cosmic Rays

In spite of all the obstacles noted above, some specific sources of cosmic-ray particles can be certified with confidence. Among these we list (1) the Sun, (2) the solar wind interacting with planetary magnetic fields, (3) the interstellar gas, (4) supernovae, and (5) radio galaxies. Needless to say, if the Sun is a source, more active stars (though farther away and not individually identified as sources) can be assured to be even more potent generators of cosmic rays. It also follows that ordinary galaxies, though not as dramatic as the radio galaxies, are also pouring high-energy particles into intergalactic space, and that the nuclei of galaxies are especially densely populated with fast particles. Furthermore, since a pulsing neutron star has been identified at the heart of the Crab Nebula, we may subdivide supernovae into two types of source: the neutron star and the nebula around it. However, the five categories of sources that have been named will serve to bring out the nature of the evidence.

1.3.   The Sun

It is paradoxical that the Sun was the first hypothesized source of cosmic rays to be eliminated from contention; yet it is the only one that has been directly identified as a source by indisputable detection of the fast particles themselves.

The early rejection of the Sun as a source was due to the smallness of the diurnal variation of counting rates at sea level. In fact, what little variation exists is now understood as an asymmetry of the radiation from remote (galactic) sources, associated with the streaming of these particles in the outflowing solar wind, rather than as a particle flux of solar origin.

However, the completeness of the lack of evidence for solar cosmic rays was heavily influenced by the use of charged-particle detectors at sea-level locations. Primary protons of kinetic energy less than 4 GeV have practically zero influence on the sea-level charged-particle flux: witness the lack of a sea-level latitude effect at latitudes below 40°. Even the most exceptional solar emission events yield extremely steep spectra, and it is very rarely (at average intervals of several years) that solar flares produce observable increments in the charged-particle flux at sea level.

The positive identification of the Sun as a source rests on detailed observations of sporadic, great enhancements of the solar particle production in events known as solar particle flares.

These were first observed during the 1940's with ionization chambers at sea level, and later with neutron monitors. Thus, the particle energy can attain 20 GeV on rare occasions. However, during the years 1942–56 only five such events were observed, giving a misleading impression of the frequency of solar emission occurrences. In more recent years other techniques have been applied that are sensitive to particles of lower energy. On the one hand, the use of balloons, rockets, and satellites has made possible detection of the particles before they are absorbed in the atmosphere, and also detection of x-rays produced by low-energy electrons entering the atmosphere. Secondly, several ground-based radio techniques have been applied to sense changes in ion and electron density in the ionosphere, caused by low-energy incoming protons. These techniques include the *riometer* (radio ionospheric opacity meter), which reveals decrements in reception of radio waves from space; the *ionosonde*, which measures radar reflection from the ionosphere; and the *forwards scatter* technique, that detects enhancements in distant reception from high-frequency radio stations. With these means, it is observed that many solar emission events are detectable each year, on the order of one per month.

The variety and complexity of the events are great: the peak intensity, energy spectrum, particle composition, anisotropy of the flux, and the intensity–time curve all vary widely from one event to another. There are many more small events than large ones. In the most energetic event prior to 1967, one that occurred in 1960, the proton flux above 20 MeV reached 3×10^4 per cm²-sec, more than 10000 times the normal average flux; and the flux integrated over the duration of the event was 4×10^9 particles per cm².‡ In contrast, smaller events are now detectable down to flux levels less than one proton per cm²-sec.

There is a strong association of the occurrence of particle flares with optical and radio flares, with the emission center usually located on the west side of the solar disk. The latter tendency can be understood in terms of the "gardenhose" field configuration in interplanetary space and the guidance of the particles preferentially along field lines. There is also a strong correlation with the solar cycle as revealed by sunspot numbers. Both the frequency and mean size of solar particle events vary in the cycle: including very small as well as large flares, the integrated number of protons above 20 MeV changes by as much as a factor 10^5 in the solar cycle.

A typical solar flare reveals the following associated phenomena:

(1) *Light.* A bright spot appears, typically extending over about 10^{-3} of the solar surface, and sometimes equalling the rest of the surface in luminosity. The flare location is in the lower corona, above a group of sunspots.

(2) *Ultraviolet.* The *U–V* enhancement exceeds that of the visible light.

(3) *X-rays.* A strong enhancement of x-rays is observed.

(4) *Radio emissions.* Prompt bursts of radio waves occur, having various frequency–time characteristics. The type IV characteristica is associated with the largest flares and suggests an emission center moving upwards in the solar corona.

(5) *High-energy particles* begin arriving at the Earth with a sudden onset that may be only a few minutes, or as long as half an hour, after the visual flare is seen. The prompt particles are very anisotropic in distribution,

---

‡ This integrated flux equals the number of galactic cosmic rays reaching the Earth in two-thirds of a century! But the average kinetic energy of the galactic cosmic-ray nuclei is hundreds of times greater than the mean energy of the solar particles, and such large flares probably occur less than once per decade.

and strike certain *impact zones* on the Earth, determined by the Earth's field and the interplanetary field configuration. These are the particles that have been guided directly (without scattering) from the Sun.

(6) *Delayed particles.* As time goes on, fast particles continue to arrive but the spectrum gets softer, the anisotropy becomes much smaller, and the particles are no longer restricted to the impact zones. These are particles that have been scattered and diffused in interplanetary space. If the flare location is not on the west limb of the Sun, the sudden anisotropic particle onset described in the paragraph above may be missed, and only the delayed particles, with a gradual onset, may be seen. The peak intensity occurs within a few hours, and the subsequent decrease follows approximately a $t^{-3/2}$ law at first, an exponential later on.

(7) *Particles in space.* Both the sudden anisotropic flux and the delayed particles have been detected at rockets and satellites in interplanetary space as well as on the Earth, with excellent correlation.

(8) *Ionization in upper atmosphere.* Only the most intense flares field sufficient prompt particle flux to ionize the atmosphere heavily in the yrst half-hour following the flare. But about a day later, the plasma emitted at the time of the flare arrives at the Earth, and with it an intense cloud of trapped particles of moderate energy (e.g., protons of 10–50 MeV). These produce ionization in the D layer (60–80 km altitude), detectable by enhanced forward scattering, by radiosonde (reflections), or by riometers as PCA (polar cap absorption) events, an absorption of cosmic radio noise. The more energetic particles, particularly the prompt ones, can produce ionization at lower atmospheric levels, resulting in radio blackouts in connection with the strongest flares.

(9) *Magnetic storm.* With the arrival of the plasma cloud, magnetometers can detect a sudden small change (SC: sudden commencement) in the horizontal component of the Earth's magnetic field (understandable as a compression of the field under the ram pressure of the plasma stream), followed by a period of magnetic disturbance that typically takes a few days to recover.

(10) *Aurora.* Displays of aurora often occur in association with the magnetic disturbance, suggesting that instabilities in the outer part of the Earth's field have triggered the release of many trapped particles from the radiation belts into the atmosphere.

This elaborate parade of phenomena sets before us a remarkable display of the intricate dynamics associated with cosmic-ray production in a small-scale outburst. With imagination one can extrapolate from this detailed, close-up view to a visualization of what happens in the explosive ejection of a large part of a whole star when a supernova occurs, and even further to the fantastic violence accompanying the explosion of a galactic nucleus or a whole galaxy.

The Sun, after all, is an extremely weak source of energetic particles. Whereas the optical radiation at the Earth from the Sun exceeds the total from all other stars by a factor of 10^{10}, the high-energy particles from the Sun, averaged over time, constitute only about one percent of the cosmic radiation. (The hard x-radiation, above 5 keV, is also less from the Sun than from other sources, in spite of its closeness.) In years when the Sun is most active, the Earth actually receives *less* particle flux than in the quiet years of the Sun, because the increased modulation of galactic radiation more than compensates for the solar emission.

Indeed, if the Sun were one of the more active sources of high-energy particles, the radiation would prevent life from existing this close to it. But being this close to a low-power source has many advantages. From more distant sources, the arrival of particles gets smeared out in time and direction, and the particles from different sources become mixed. The result is that individual sources cannot be distinguished and the detailed history of production events cannot be followed. Thus, the galactic cosmic rays give a false impression of simplicity, regularity and constancy in the production process. The Sun is the only place where the particle history of events can be resolved and studied in detail. They may not be qualitatively identical to the processes occurring in the major, more remote sources, but some resemblances are likely, and it is the most high-power laboratory we can use.

1.4.   Artificial Plasmas

The Sun is not as accessible as a man-made laboratory, and even though the latter cannot approach astrophysical conditions very closely, the laboratory experiments motivated by fusion power development yield relevant observations. It is consistently revealed in the "plasma pinch" process and other examples of plasma instability that whenever turbulent motion develops in a plasma, high-energy particles arise. That is, the agitation of plasma results in a small fraction of the particles being accelerated to energies greatly exceeding the average thermal value.

1.5.   Interplanetary and Planetary Sources

On an intermediate scale, particle acceleration has been observed to occur both in interplanetary space and in the region of interaction of the solar wind with the Earth's magnetic field. The remarkable radio waves from the belts around Jupiter inform us that the Earth is not unique, but similar acceleration processes occur in even greater intensity around the larger planet.

In the interplanetary region, a solar blast wave sometimes plows through the medium at shock velocities exceeding 1000 km/sec, piling up the normal, slower wind ahead of it and causing disordered turbulence near the head of the wave. An instance recorded on Explorer XII was reported by Bryant *et al.* (1962), in which the proton flux between 9 and 14 MeV underwent a sudden tenfold increase when a solar blast wave neared the satellite, and diminished soon after the head of the stream had passed.

In the frame of reference of the solar wind plasma, the Earth is moving at a speed greatly exceeding the Alfvén wave velocity; hence a magneto-hydrodynamic shock wave forms on the subsolar side of the Earth. It is not the Earth itself or its atmosphere that pushes aside the solar plasma, but the Earth's magnetic field, so it is a collisionless shock, at about $13.4R_E$ (Earth-radii) from the center. Between the detached bow shock at $13.4R_E$ and the magnetopause at $10R_E$ (where the magnetosphere terminates on the subsolar side of the Earth), a region of disordered fields and turbulent motion exists called the magnetosheath. Examples of acceleration in this region and upstream from it are in the report by Steljes *et al.* (1961) on protons associated with a solar blast wave, and the reports of electron acceleration by Fan *et al.* (1964, 1965), K. Anderson (1965), and Frank *et al.* (1963).

On the midnight side, the Earth's field lines are stretched out by solar wind, creating a turbulent wake that extends at least as far as the Moon. Between the oppositely directed field lines of the northern and southern hemisphere a "neutral sheet" forms. Here the continual annihilation of field lines (replaced by others moving towards the equator) transfers magnetic-field energy to particles, the solar wind being the ultimate source of the energy (Speiser, 1965; Murayama and Simpson, 1968).

By these means and possibly others, large densities of fast particles get produced and trapped in the van Allen belts around the Earth and Jupiter. Around the Earth, typical electron fluxes are 3×10^7 per cm²-sec above 40 keV, 5×10^6 above 500 keV, while the proton flux above 40 MeV (in the inner belt) is about 2×10^4 per cm²-sec, with a differential spectrum proportional to $E^{-1.8}$. The decay of albedo neutrons (produced by galactic cosmic rays in the atmosphere) is insufficient to account either for the electrons or for the low-energy protons, especially the large numbers below one MeV: moreover, the spectra have the wrong shape to be identified with neutron albedo origin. The principal energy apparently comes from the solar wind, both in its steady and gusty aspects.

1.6.   Interstellar Space as a Source

We turn now to brief consideration of the interstellar gas as a source, leaving more detailed considerations to later sections on the origin of positrons, electrons, x-rays, and gamma rays.

The interstellar space was once proposed by Fermi (1949) as the location of the principal cosmic-ray acceleration. The acceleration occurring here is no longer believed to be dominant in the production of cosmic-ray energies, but it certainly occurs to some degree, owing to the existence of ionized gas clouds in random motion, carrying field lines that are frozen in the plasma because of its high conductivity. Cosmic rays in random motion are reflected by such clouds, with an average gain in energy of $8\beta_0^2 E/3$ per encounter, where E is the original particle energy and $\beta_0 c$ the velocity of the clouds. Thus the energy rises exponentially with time according to $E = E_0 \exp(8\beta_0^2 t/3t_1)$, where t_1^{-1} is the mean frequency of collisions. It is presumed that either a nuclear collision or diffusion out of the Galaxy terminates the process after a mean time t_2, with the probability of survival going as e^{-t/t_2}. By joining these expressions one obtains a predicted power-law energy spectrum with an exponent (in the integral spectrum) equal to $-[3t_1/(8t_2\beta_0^2)]$, if particles are continually injected at one energy.

The trouble with this model is that the rate of energy gain is too low: realistic values of β_0, t_1 and t_2 yield much too steep a spectrum. One cannot presume a small value of t_1, or the clouds could not be large enough to reflect high-energy particles; and one cannot postulate a very large value of t_2, or the heavy nuclei in the cosmic radiation would all be broken up in nuclear collisions before attaining high energy; while β_0 is limited by observed motions of interstellar clouds. The reason for mentioning these considerations is the possibility that in regions more active than the average interstellar space (e.g., in a galactic nucleus), more frequent encounters may occur with more concentrated gas clouds in more violent motion, and the Fermi process may then be rapid enough to be the dominant accelerator.

The formulas given above assumed random cloud motions. If a particle is trapped in a magnetic bottle between two approaching clouds, the angles of successive encounters are not random, and the fractional energy gain per collision is of order β_0 instead of β_0^2, an enormous gain if $\beta_0 \lesssim 10^{-4}$. The acceleration then proceeds rapidly until the particle acquires enough rigidity to penetrate one of the clouds. However, this process requires such special conditions of motion and of field configuration between the clouds that it would very rarely occur.

Even though most of interstellar space is not very effective in accelerating the nuclei which are the major constituents of cosmic rays, interactions certainly occur in this space that make it a source of other components of the cosmic rays—components that may not be produced as abundantly at the primary cosmic-ray sources.

For instance, electrons are produced by the knock-on process. As discussed in more detail in a later section, this process in interstellar space seems to be the dominant source of the non-solar electrons under about 20 MeV. Both positrons, electrons, and neutinos are produced in decay processes following the production of pions in nuclear collisions: this is apparently the main source of positrons and high-energy neutrinos (other than the secondary positrons and neutrinos produced by cosmic rays in our own atmosphere). A small flux of antiprotons also arises from these collisions, though it has not yet been detected. High-energy photons are produced in the decay of neutral pions. It is not yet clear whether or not this is the principal source of the primary gamma rays above 10 MeV: this will need discussion in a later section.

Some nuclear species, although stable against decay in a vacuum, are very rare in stellar matter because of large cross sections for nuclear reactions, and low values of the binding energy in comparison with neighboring nuclides. These rare species include particularly the nuclei of elements 3, 4, and 5 (lithium, beryllium, and boron), which are extremely rare in stars. Deuterium is also rare, and helium 3 much less abundant than helium 4. In general, odd elements such as fluorine are rare compared with neighboring even elements like oxygen and neon. Thus, the material directly accelerated should have striking deficiencies of numerous nuclear species. However, when a heavy cosmic-ray nucleus encounters a proton of the interstellar gas, the cross sections for production of those rare nuclei by a spallation process are not very low. The primary cosmic rays apparently diffuse in the Galaxy for a million years or more, on the average, before reaching the Earth or escaping, and in this time a significant number of the nuclei undergo nuclear collisions with the gas. Such collisions are therefore a major source of the rarer nuclei.

One of the difficulties in interpreting the data on this subject is that the cosmic rays may also spend a significant time diffusing in their sources before escaping into interstellar space; so not all of the fragmentation needs to have occurred in the latter region. However, the degree of fragmentation of heavy nuclei and production of rare isotopes at least sets an upper limit on the time of diffusion in the Galaxy.

Besides interacting with the gas, the cosmic rays in interstellar space can interact with the magnetic fields and with the background radiation (microwave, infrared, and visible). These electromagnetic interactions

2*

are generally important only for the electron–positron component of cosmic rays (except at extremely high energy), because the mass of the particle occurs to the inverse fourth power in the cross sections. Electrons interacting with the magnetic field produce synchrotron radiation of predominantly radio frequency; but upon scattering off the background radiation, they produce hard x-rays and gamma rays. The latter process is competitive with the production of gammas by pion decay, and occurs with sufficient frequency to be expected to modify the high-energy part of the electron spectrum.

Very high energy cosmic gamma rays interact with background photons to make electron–positron pairs. The energy requirement is that the product of the energies of the two photons exceed the square of the electron rest-energy. This process is not an important source of electrons and positrons, but serves to make the Universe opaque to photons above about 10^{14} eV. Ultrahigh-energy protons can undergo photonuclear reactions with the background radiation resulting in pion production, when the product of proton and photon energies exceeds the product of pion and proton rest energies (about 10^{17} eV^2). This process not only shortens the lifetime of protons above 10^{20} eV, but is probably the dominant (though very weak) source of photons and neutrinos at energies around 10^{19} eV.

1.7.   Supernovae

At this point we do not want to engage in speculation about the detailed mechanism of generating high-energy particles in supernovae, but only to mention the evidence that the supernovae do, in fact, constitute potent sources of cosmic rays. This evidence lies primarily in the special properties of the intense radio, infrared, visible, and x-radiation from these sources. The shape of the spectrum and the polarization of the radiation confirm that at least up to the ultraviolet, and probably up to the x-ray frequencies, the emission mechanism is magnetic bremsstrahlung by high-energy electrons. The source is distributed throughout the nebula, and limitations of the magnetic-field strength compel one to conclude that many of the electrons have energies on the order of 10^{12} eV (or even higher if the x-rays are established to be emitted by the same mechanism). The radiative emission by these electrons is proceeding at a rate of about 10^{38} ergs per second and lasts for thousands of years, giving a total yield on the order of 10^{49} ergs. From the boundary of a supernova, many of the high-energy electrons must also escape into interstellar space, particularly as the envelope expands and the fields weaken.

Direct detection of the presence of very high energy protons in supernovae has not yet been possible, mainly because heavy particles are such poor radiators. But it is hard to conceive of a mechanism that would accelerate the electrons in the nebula without accelerating protons also. The most plausible mechanisms, in fact, accelerate the nuclei to higher energies than the electrons; and the nuclei retain their energy instead of losing it in radiative processes, so a larger part of this energy is released into interstellar space. Among the cosmic rays generally, nuclei are more abundant than electrons by a factor of 100. If the total output of energetic nuclei per supernova is 10^{50}–10^{51} ergs, the input to the Galaxy is about 10^{41} ergs per second, since supernovae occur about once per 100 years in the Galaxy. This energy input is sufficient to maintain the observed cosmic-ray flux.

Moreover, the spectrum of the synchrotron radiation from supernovae, though somewhat variable from one to another, implies a particle energy spectrum similar to that which is observed in the cosmic rays. And the cosmic rays, like the material of supernovae, are rich in heavy elements, relative to the composition of other stars. Thus it appears that supernovae may be the principal source of cosmic rays in the Galaxy.

It has now been established that a neutron-star pulsar resides at the center of a supernova, and its regular pulsation is certainly associated with rotation. The period is so precise that the gradual slowing of the rotation due to energy dissipation has been measurable (Fishman *et al.*, 1969; Richards and Comella, 1969; Warner *et al.*, 1969); the rate of rotational energy loss is adequate to maintain the supply of high-energy electrons in the nebula and provide the energy of the cosmic-ray nuclei. The explosive expulsion of the envelope of the star, associated with collapse of the core to form the neutron star, would also have provided an initial supply of high-energy particles (Colgate and White, 1966). Thus, an adequate driving mechanism for the supernova has been identified. Plausible models for the detailed energy transfer from stellar rotation to high-energy particles have also been proposed, but we will not discuss these here.

1.8.   Radio Galaxies and Quasars

Radio galaxies are identified as massive cosmic-ray generators by evidence similar to that adduced for supernovae. The radio emission is certainly by the synchrotron process, and in at least one case (the jet in Virgo A), the visible light has been found to be polarized and therefore is also of magnetic bremsstrahlung origin. The Virgo A galaxy, interestingly, is also a tremendous source of hard x-rays and gamma rays, with a spectrum that is a direct extension of the radio and optical spectrum, suggesting that even the x and gamma rays may be of synchrotron origin.

The total radiation from these huge, violent structures exceeds that of one supernova by factors ranging from 10^4 to 10^8. There is no doubt that the processes which we recognize as productive of cosmic rays in our Galaxy are occurring on a greatly multiplied scale in the radio galaxies.

There is some question, however, as to whether the particles accelerated in remote galaxies ever reach the Earth. This question is not easily answered and is beset with the uncertainties attending all inherently cosmological problems. There is also doubt as to how open is the field structure of a galaxy: that is, what is the time scale for escape of its cosmic-ray product into space. It appears that active galaxies represent a comparatively brief phase (on the order of 10^6 years) in the evolution of many galaxies. On this basis, estimates have been made of the input into intergalactic space, with the result that the cosmic-ray flux there may be on the order of 10^{-3} times that in our galaxy. Of course, this estimate may be wrong by a factor of 100 either way.

We shall return to the above question later in discussing the isotropic part of the x-ray and gamma-ray flux, for this evidence can set limits on the number of energetic particles in the Universe. At this point let it suffice to say that the background x-rays and gamma rays are easier to account for if one assumes fast electrons to be present in intergalactic space, with a spectrum that has been steepened by long exposure to radiative energy losses, but which, without these losses, would be about 10^{-2} of the electron flux in the Galaxy.

Indirect evidence for an extragalactic component of the cosmic rays is also provided by changes in the slope of the spectrum at very high energies, and apparent changes in the primary composition. From 10^{10} eV to 10^{15} eV, the shape of the spectrum is well represented by $dJ \sim E^{-2.6} \, dE$, and the relative proportions of heavy and light nuclei are essentially constant. Between 10^{15} and 10^{18} eV, the spectrum steepens to about $E^{-3.2}$ in form, while beyond 10^{18} eV it appears to resume its previous slope (Fig. 1). Moreover, the heavy nuclei seem to have disappeared at the higher energies (Linsley, 1962). The implication seems to be that galactic retention of cosmic rays weakens above 10^{15} eV per nucleon, and that the nuclei above a total energy of 10^{18} eV represent those that pervade intergalactic space, at least within the local cluster.

The extrapolation of the lower-energy part of the spectrum lies above the high-energy branch by a factor of about 50, suggesting that the extragalactic cosmic-ray spectrum is roughly parallel to that within the

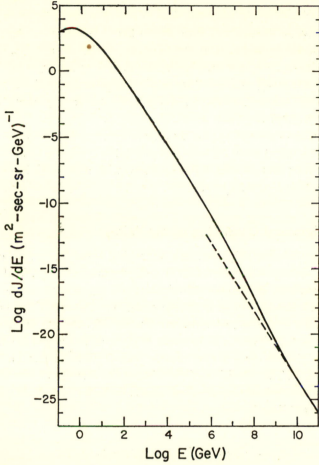

FIG. 1. Total primary differential spectrum at the Earth in a year of minimal solar activity. In the lowest decade of kinetic energy, the interstellar flux is higher: the curve has no maximum. The dashed line at high energies may possibly represent the intergalactic flux.

Galaxy but about two orders of magnitude weaker. The absence of heavy nuclei is likely to apply only at very high energies, where heavy nuclei can be depleted by photonuclear reactions with starlight. Particles of lower energy, diffusing outside the Galaxy, can leak in as well as those inside can leak out. Thus we see a likelihood that one in a hundred (or one in ten, or one in ten thousand) of the cosmic rays we detect may have come from a remote galaxy. If half of the galaxies are composed of antimatter, there should be a realistic chance of detecting such things as anticarbon nuclei, which can *not* be made by collisions of ordinary nuclei in interstellar space.

2.   ESP (ENERGETIC SOLAR PARTICLES)

Following the general summary in Sec. 1, we turn in the present section to a more detailed look at the character of the particle flux of solar origin. In doing so, we lean heavily on the review article by Carl Fichtel and Frank McDonald, entitled "Energetic Particles from the Sun" and published in the Annual Reviews of Astronomy and Astrophysics (Fichtel and McDonald, 1967).

As mentioned before, the Sun is something like a volcano, blowing out gas in a great wind all the time, though varying in force with the eleven-year cycle; and sporadically exploding with abrupt violence in the discrete occurrences called flares.

2.1.   Origin and Character of the Wind

The solar wind originates in a hot region at the base of the corona, that probably derives its temperature from hydromagnetic waves of deeper origin. The distribution of states of ionization of oxygen has been measured, with the result that the most probable states are OVI and OVII, i.e., that the temperature at the source is about 2×10^6 °K. As the wind rises in the solar atmosphere it increases in speed because of the continued pressure gradient, diminishing density and reduced gravitational field; thus it becomes supersonic. The condition of the wind at the Earth, of course, varies; but may be characterized in quiet solar times and times of solar disturbance by the entries in Table 1.

TABLE I.   Conditions in the solar wind at one AU.

	Times of quiet sun	Times of disturbed sun
Atomic density, n (cm^{-3})	3–10	20–40
Bulk velocity (km/sec)	300–400	800
Alfvén speed $[B/(4\pi\varrho)^{1/2}]$ in the solar plasma (km/sec)	50	100
Magnetic flux density, B (gauss)	$(3–8) \times 10^{-5}$	$(10–30) \times 10^{-5}$
Proton energy at bulk velocity	800 eV	4 keV
Proton gyroradius	1000 km	1000 km
Energy density of field (eV/cm^3)	60	1000
Thermal energy density (eV/cm^3)	60	1000
Energy density in bulk gas motion (eV/cm^3)	8000	100000
Energy flux (ergs/cm^2 sec)	0.5	15

The magnetic field lines in quiet solar times are oriented generally outwards or inwards along spiral curves originating at the Sun, and divided in polarity around the equatorial band into four great sectors. Of course, there are local irregularities superimposed on this regular structure, and they are accentuated in disturbed periods. The normal spiral configuration is not due to azimuthal motion of the plasma (in fact, its motion is radial), but to the lines of force being frozen in the highly conductive plasma and rooted in the rotating Sun.

Because of the conductivity of the plasma, the Earth creates a big geomagnetic cavity in the outflowing wind: during the short time in which the wind sweeps by the magnetosphere (about 3 minutes for a distance of ten Earth radii), the Earth's field cannot leak far into the stream. Indeed, a long wake is created on the night side of the Earth, extending on the order of 100 Earth radii.

A solar flare originates in the corona over an active sunspot group. Magnetic fields control its geometry and history. Dark filaments are observed to brighten suddenly. When seen at the solar edge, these filaments form loop prominences that become highly luminous and expand upward. The optical rise time is 5 to 20 minutes and the duration on the order of 20–90 minutes.

A blast wave is then emitted, or plasma stream that travels faster than the normal solar wind and creates a "bump" in the transverse field by compressing the region ahead of the shock front (see Fig. 2). When such

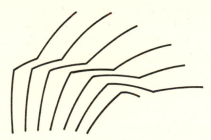

FIG. 2. Illustration of distortion of field lines and generation of a transverse magnetic pulse by a gust in the solar wind.

a front arrives at the Earth, about two days after the flare on the Sun, it initiates a geomagnetic disturbance, followed by a "Forbush decrease" in the galactic cosmic-ray flux—a sudden drop followed by a long period of slow recovery. The drop in cosmic rays is due to their not having had time to diffuse into the denser and faster-moving plasma behind the shock front as thoroughly as they have leaked into the weaker field and slower-moving plasma ahead of the front. This effect is not limited to low-energy particles, but is also evident in primaries at tens of GeV energy, which produce the muons reaching sea level.

However, the arrival of the blast wave may bring about an increase in the flux of comparatively low energy particles in at least two ways: by transport from the Sun, and by acceleration in the agitated fields near the shock front.

2.2.   Time Profiles of ESP

The energetic particles from the Sun (ESP) have linear speeds far in excess of the wind velocity. For instance, a proton of 30 MeV kinetic energy has $v/c \cong 1/4$. Therefore in a favorable guiding field these particles and the ones of still higher energy are capable of arriving with very little delay—half an hour or less—following the visual flare; but they do not always do so. The time profile of an event has numerous features: time of onset, rise time, fine structure, and time of decay. There are also changes with time in the directional asymmetry and the energy spectrum. All of these features are variable from one event to another. In seeking to understand these things, it is well to remember that the proton of 30 MeV has a rigidity of about 250 MV and thus a very short gyro-radius, on the order of 3×10^9 cm, in the fields carried by the blast wave. Moreover, the ESP are not numerous enough to alter the fields. Very large flares may yield a flux of ESP on the order of 1000 per cm²-sec, but this implies a particle density $n \cong 10^{-7}$ cm⁻³, about 10^{-8} times the gas density, and an energy density on the order of 10^{-4} times that in the gas stream: so it is the more numerous particles of low energy and the fields they transport, as well as the pre-existing fields in interplanetary space, that determine the geometry of the events.

Processes that are essential in accounting for the time profiles of ESP include (1) guidance by interplanetary fields, (2) storage in magnetic trapping regions of several types, and (3) scattering by magnetic irregularities. The combination of guidance with scattering gives rise to *anisotropic diffusion*, a particle drift with much larger diffusion coefficient along the field than transverse to it.

When a flare occurs in a favorable location, near the west limb of the Sun, the fastest particles arrive at Earth abruptly and very asymmetrically. The delay in onset is due to transit time only, and the rise time, an hour or less, is similar to that of the optical flare. The direction of the maximum flux approaching the Earth is close to that of the spiral field configuration. Guidance by the pre-existing interplanetary field clearly dominates the geometry in first approximation and the early time distribution is explainable in terms of the variation in pitch angles. The most energetic particles not only travel fastest but are obviously accelerated and released from the Sun quickly. The extreme directionality of the particles at early times in the flare is influenced by the fact that with increasing distance from the Sun, the guiding field gets weaker and the transverse particle momentum decreases: $p_\perp{}^2/H \cong$ constant; thus the pitch angles get small.

When a flare occurs in a less favorable location on the Sun, the onset of ESP flux at the Earth is more gradual, the peak flux tends to be less, and the rise time is longer, sometimes many hours. In prompt flare events yielding energetic electrons, the source of radio noise is usually well localized near the Sun, while in nonprompt electron events the radio source is widely distributed over the solar disk, suggesting the trapping of particles at the Sun and diffusion over the surface as well as upwards in the corona.

The decay of a prompt particle flare begins as a $t^{-3/2}$ power law and this is followed by exponential decay, extending over a day or several days.

The strong directional asymmetry of prompt flares is lacking in those of gradual onset, and in any case disappears by the time the exponential decay sets in. The loss of directionality with time, the gradual onset of the non-prompt flares, and the slow exponential decay are accounted for by scattering and ultimate leakage from the system.

The energy spectrum of the particles also varies with time and differs from one flare to another. High particle energies are more common at early times in prompt flares: shortly after injection the spectrum resembles a power law in momentum, albeit with a rather large exponent compared with that of galactic cosmic rays. In the decay phase, however, the spectrum is more nearly exponential, going as e^{-R/R_0} in rigidity R, with the value R_0 less than 400 MV and diminishing with time. In this phase, therefore, there are very few particles of high energy.

Energetic solar particles are not always released to travel ahead of the blast wave to the Earth. Not only can they be trapped at the Sun, but they can be contained by the advancing bubble of plasma and transported with the gas streams. Such streams may then encounter a satellite or the Earth, or may drift across such a body without meeting it head-on. In observations made on space vehicles a filamentary structure of such particle streams has been evident, with extremely sharp boundaries, giving rise to large and rapid oscillations in the detected particle flux.

There are also *recurrence events* in the solar particle streams, particularly in the case of rather low energy particles. That is, repetitions of particle enhancement occur at intervals of 27 days, the solar rotation period. The explanation requires either a long-term storage at the Sun or a continued acceleration process extending over months of time. As mentioned above, there is other evidence of trapping at the Sun, partly in the spreading of radio and particle source regions over the surface, and partly in the fact that some optical flares yield no detectable energetic particles at all.

2.3.   Energy of a Flare; Integrated Solar Energetic Particle Flux

A large solar particle flare produces an integrated particle energy flux at the Earth on the order of 10^4 ergs per cm^2: the largest flare recorded gave an order of magnitude more than this. These numbers are impressive compared with the average galactic particle flux, which is only about 4×10^5 ergs across one cm^2 of the Earth's atmosphere per *year*. However, the large flares are very infrequent, and adding the energy flux from smaller flares does not raise the total greatly. An effort has been made to record extremely small particle flares, called "micro-events", but even though these are very frequent (they are going on almost all the time), the particle energies are so low and the particle flux so small that their integrated energy is insignificant. In short, the total ESP energy flux per year at the Earth is about 10^5 ergs per cm^2.

Comparing this directly with the galactic flux at the Earth is not very meaningful, because the solar particles are mostly below 100 MeV in energy, and the bulk of the galactic flux below 100 MeV is prevented from reaching Earth by the solar wind, and has not been included in the galactic energy flux estimated above.

The total output from the Sun in ESP is about 10^{25} ergs per second on the average, which is very little compared with the thermal luminosity, 4×10^{33} ergs/sec, or even compared with the solar wind, which involves a power output of 10^{27} ergs/sec for the quiet Sun, and 10^{28} ergs/sec in active years. If all the stars of the Galaxy produced energetic particles as the Sun does, the cosmic-ray input to the Galaxy would be 10^{36} ergs per second, far short of the 10^{41} ergs/sec needed to maintain the observed flux. In addition, the solar spectrum is much too soft: it hardly ever produces any particles above 10 GeV, the average energy of the galactic radiation.

Nevertheless, even at the Sun, it is impressive to note the high efficiency in nature's careless, off-hand way of running an accelerator. A large solar flare yields an optical energy release of about 10^{32} ergs and a blast wave carrying a kinetic energy also of 10^{32} ergs, while the energy output in ESP is about 10^{30} ergs. The most carefully engineered, man-made accelerators only have an efficiency of 10^{-3}, which is apparently not quite as good as that of an uncontrolled natural explosion, provided the latter occurs on an astronomical scale.

2.4.   Composition of ESP

Table II gives the approximate relative numbers of various nuclear species among the energetic solar particles, together with corresponding numbers for the galactic cosmic rays. The comparisons are made at equal energies per nucleon, and the normalization is to oxygen. The figures in the table are only approximate, because the available data are not in complete agreement, and the composition apparently varies slightly with energy.

TABLE II.   Cosmic-ray composition.

Element	Solar particles	Galactic cosmic rays
H	~ 1100	500
He	100	50
Li, Be, B	< 0.02	0.8
C	0.6	1.4
N	0.2	0.3
O	1.0	1.0 (normalized)
F	< 0.03	< 0.02
Ne	0.13	0.2
Na	< 0.03	0.1
Mg	0.04	0.3
Al	< 0.02	0.1
Si	0.03	0.2
15–21	0.06	0.1
22–28	< 0.02	0.2

A general observation is that the composition of the solar particles is like that of the photosphere, rather than like that of the corona: i.e., the abundance of the medium-weight nuclei is moderately high. Thus, although the radio bursts are emitted in the corona, the accelerated particles are drawn from the photosphere.

There are less heavy elements such as iron in the solar than in the galactic particles. This is consistent with the belief that many of the galactic cosmic rays come from supernovae, which represent a later stage of element synthesis than does a main-sequence star.

The solar particles show very little evidence of elements that would only arise by fragmentation. No solar neutrons have yet been detected. Deuterium and He^3 are present in very small numbers; deuterium about 10^{-3} times as abundant as protons. Lithium, beryllium, and boron are very rare. Therefore the solar particles spend very little time after acceleration in a region of appreciable density: the matter traversed is less than 0.1 g/cm^2. Taking the density at the source to be 10^9 atoms per cm^3, the storage time can only be about an hour. There is evidence, discussed above, of much longer storage than this on some occasions; therefore the accelerating process must draw the particles up into the corona where the density is less.

The relative abundances of the heavy nuclei are constant within the accuracy of available measurements, but the ratio of hydrogen to helium varies substantially. This is not surprising, since the heavier nuclei all have nearly equal charge-to-mass ratios, while hydrogen has twice this charge per unit mass; therefore the magnetic rigidity of protons is lower at a given velocity than it is for the other nuclei.

A particularly strong variation, however, is found in the ratio of electrons to heavy particles. Given the long storage and thorough mixing of particles from other cosmic-ray sources, there is little chance of detecting such a variation in composition from any source but the Sun. Here it is extremely interesting, because it suggests that the mechanism for accelerating or releasing electrons may *not* be closely coupled with the corresponding mechanism for protons and heavy nuclei.

On the average, the solar electrons are about as numerous as the protons, but the methods of detection that have been applied are sensitive to electrons down to a few keV in energy, while protons are usually only registered if their energies are several MeV or more. Thus, the *energy* in the electron component is normally much less than that carried by the protons. Of course, electrons can also lose energy by radiative processes in the coronal fields, to which the protons are impervious. Some flares, however, yield many electrons with no detectable protons at all. Furthermore, the timing is different; often the electron component arrives hours before the protons from a slow-onset event.

Thus, solar flares are highly variable in particle composition as well as in intensity and time profile. It remains to be determined whether basically different flare and accelerating processes occur, or whether all this variability can be accounted for in terms of incidental geometrical differences between flares, and variations in their degree of violence, together with the big difference in charge-to-mass ratio between the electrons and atomic nuclei.

3.   DISCRETE SOURCES REMOTE FROM THE SUN

The Sun, the planetary magnetospheres, and interplanetary space have been positively identified as sources of energetic particles by means of sharp changes of particle flux with time, the striking anisotropy of the flux on some occasions, and the detailed correlation of particle intensity with other phenomena (principally optical and radio signals) that have temporal and directional features of even better definition. The energetic particles produced by these sources have thus been distinguishable from the more continuous background flux, and have been directly measurable (though they are extremely variable) in number, energy spectrum, and composition.

Remote sources, on the contrary, have had to be inferred indirectly, because the temporal and directional features of the particle flux from such sources are completely washed out in the process of interstellar diffusion. Furthermore, the particles lose energy and suffer fragmentation in space, and particles that were not present in the sources are produced in the interstellar medium. Finally, the solar wind drastically modulates the low-energy part of the flux from remote sources as it enters the interplanetary region. These effects cause serious uncertainties in any conclusions reached about quantitative details of the production by particular objects. Nevertheless, the fitting together of many kinds of evidence gives confidence in the identification of some places as extraordinarily potent sources of high-energy particles. Dominant among those in the Galaxy are the supernovae, while externally there are radio galaxies, quasars, Seyfert galaxies, and exploding nuclei in galaxies of other classifications.

3.1.   Supernovae

The distances to the nearest supernovae (SN) that are still active measure thousands of light years. The mean propagation velocity of particles, relative to that of light (assuming the particles to be relativistic) is λ/R, where λ is the mean free path for random scattering, and R the distance from the source. The isotropy of cosmic rays requires that λ be no more than a few light years. Hence, the propagation time is millions of years. By the time that an appreciable flux diffuses to the Earth from an SN a few thousand light years away, the stellar remnant itself is so old that it is not optically observable. Therefore, particles have not been detected at the Earth from observed SN, and are not expected to be.

At some time in the past, it is likely that one or more SN explosions occurred less than fifty light years from the Earth. Then the Earth must have been bathed in particle radiation while the nebula was still visibly active. The time constant for flux change at the Earth would have been long, but measurable by an advanced civilization; and the flux would have had a noticeable anisotropy. But it is also probable that the complex organisms which may have constituted the advanced civilization were destroyed by excessively rapid mutations induced by the radiation itself, and there would now be no clear record of its occurrence—except for geological evidence, such as we have, of the strange extinction of highly organized biological species in the remote past.

However, we can infer the presence and production of energetic particles *in* the SN remnants that are still observable, from the types of radiation that propagate to us at speed *c*. The assumption that similar events occurred in the past then leads to the conclusion that the galactic particles now reaching us may have come from past SN *like* the recent ones we see now.

Direct proof of energetic particles in SN exists at present only for the electrons. The evidence is in the polarized, continuous radiation in the radio and optical bands (and possibly the x-ray and gamma-ray bands as well), that has been identified as magnetic bremsstrahlung, or synchrotron radiation.

Synchrotron Radiation (Ginzburg and Syrovatskii, 1964)

Nonrelativistic particles circulate and radiate at a definite low frequency and power in a magnetic field of strength B: $v_c = Be/2\pi mc$ = 2.8 Hz *per microgauss* for electrons, and 5.5 cycles *per hour* per microgauss for protons. But highly relativistic particles with Lorentz factor γ ($\gamma = E/mc^2$) emit high harmonics of the cyclotron frequency, with a spectral distribution that peaks at $v \cong 0.34\gamma^2 v_c$. The *number* of photons emitted per unit time is proportional simply to Bz/m and is independent of γ, but the average photon energy is proportional to $(Bz/m)\,\gamma^2$. [Here ze is the charge of the particle.] Hence the power radiated is also proportional to γ^2:

$$\frac{dE}{dt} \text{ (per particle)} \propto \frac{B^2 z^2}{m^2}\left(\frac{E}{m}\right)^2.$$

Heavy particles emit at a frequency porportional to m^{-3} and a power proportional to m^{-4}, which for protons is 10^{-13} in comparison with the power radiated by electrons of the same energy. Thus, synchrotron radiation by any nuclei that are present is completely insignificant compared with that emitted by the electrons.

In simplified discussions of the characteristics of supernovae, the object is often treated as though the field strength and particle density were uniformly distributed throughout the observed volume. This is admittedly a crude approximation, and we shall consider alternatives below. Objections to the uniform model are particularly justified since the 1969 discovery of pulsed radio, optical, and x-radiation coming from a neutron star near the center of the Crab Nebula. However, more than 90 percent of the radio, optical, and x-radiation of Tau A comes from the distributed nebula rather than the central pulsar, so our simple model still has relevance.

Limits on Field Strength in Uniform Model

Large values of magnetic-field strength, B, are usually excluded on the following grounds.

(1) Absence of large Zeemann effect in the line optical radiation.

(2) Electron lifetime: the time in which an electron loses half its energy is given by $T = E/(dE/dt) \propto (EB^2)^{-1}$. The energy of the electrons principally responsible for the radiation of frequency ν is proportional to $(\nu/B)^{1/2}$; hence $T \propto \nu^{-1/2}B^{-3/2}$. Large values of B would not only preclude survival of electrons for the history of the nebula (915 years in the case of Tau A), but at the higher frequencies would not even leave time for the electrons to diffuse over the nebula; hence they would require the seat of electron acceleration to be spread throughout the nebula.

(3) Total energy. The total power radiated in an interval $d\nu$ is given by $dP \propto B^2E^2\, dN$ (omitting constants of proportion), where dN is the number of electrons responsible for the radiation in the frequency interval, and E is their energy. For a given amount of power received at Earth (an experimental observation, not a free parameter), the total electron energy at the source is therefore given by $E\, dN \propto B^{-2}E^{-1}$. However, $E \propto (\nu/B)^{1/2}$; thus $E\, dN \propto \nu^{-1/2}B^{-3/2}\, dP$. Integrated over the electron spectrum, the total electron energy is proportional to $B^{-3/2}$. Meanwhile the total energy of the magnetic field is proportional to B^2. Thus, the combined energy is

$$W = aB^2 + b/B^{3/2}.$$

This has a minimum value, $W_{\min}$, at $B_{\min} = (3b/4a)^{2/7}$. At other values of B,

$$\frac{W}{W_{\min}} = \frac{3}{7}\left(\frac{B}{B_{\min}}\right)^2 + \frac{4}{7}\left(\frac{B_{\min}}{B}\right)^{3/2}.$$

Thus, if B is imagined to be even ten times $B_{\min}$, the total energy requirement is raised by a factor of 43, and if B is supposed to be 100 times B_{min}, the total energy needed is 4300 times W_{min}. Setting $B \cong 5 \times 10^{-4}$ gauss $\cong B_{min}$ for the Crab Nebula, one finds that the magnetic energy alone is 10^{48} ergs in this nebula, and if B were 100 times higher, the magnetic energy would be 10^{52} ergs, the total energy released in the explosion (and the probable total rotational energy of the central neutron star immediately after collapse). Clearly this is too much. Not only would insufficient energy be available for such fields, but the magnetic pressure would have generated instability, accelerating the outwards flow of gas more than is observed. Thus, in a uniform model B is certainly less than 10^{-2} gauss and a more reasonable upper limit is 5 milligauss.

(4) Degree of linear polarization and smallness of circular polarization in the radio spectrum.

(a) The radiation produced at different depths within a nebula suffers different amounts of Faraday rotation, tending to depolarize the total radiation at the receiver. Comparison of the net polarization at different radiofrequencies determines the mean value, along the line of sight, of the product $n_e B_{\parallel}$ (electron density times field component along the line of sight). The fact that linear polarizations of the radiation from supernovae are several percent at wavelengths as long as 20 cm implies that the values of this product are small. For instance, Seielstad and Weiler (1968) obtain 7×10^{-5} gauss-cm^{-3} for the Crab Nebula. This cannot be due in large measure to cancellation of rotations by fields in opposite directions, because the direction of the polarization is remarkably uniform over large areas of the nebula. The electron density cannot plausibly be much less than one per cm^3; hence B cannot be very large compared with 10^{-4} gauss.

(b) Synchrotron radiation is plane polarized only in first approximation; more generally, it is elliptically polarized (Ginzburg and Syrovatskii, 1964; Legg and Westfold, 1968). The degree of circular polarization is inversely proportional to the energy of the electrons, hence it is proportional to $(B/\nu)^{1/2}$ where B is the field strength and ν the frequency. For the Crab Nebula, the circular polarization has been found to be less than 1% at 38 MHz (Andrew, Purton, and Terzian, 1967) and less than 0.2% at 1400 MHz (Seielstad and Weiler, 1968). Both measurements imply $B < 10^{-3}$ gauss. Similarly small limits have also been obtained for other supernovae. This result is independent of the Faraday depolarization and applies to the field causing the synchrotron emission, rather than to the field through which the radiation propagates after emission (which produces the depolarization).

Very weak magnetic fields can also be excluded:

(1) As the supposed field strength is reduced, the required electron

energy per particle goes up; but the field must still be able to contain the particles. If the field were as weak as one microgauss, the electron energy needed to produce the polarized blue light (4000 Å) would be 10^{13} eV and the cyclotron diameter would be 10^{17} cm, too high for retention since the observed coherence length of the field (as judged from polarization measures) is barely larger than this, and effective retention requires the cyclotron diameter to be very small compared with the dimension over which the field is coherent.

(2) Total energy. The same argument used to exclude excessively strong fields can be applied to exclude weak ones. If the field is considered to be small, the energy of all the contained electrons increases as $B^{-3/2}$, and ultimately exceeds the total energy released in the explosion. Before this stage is reached, the particle *pressure* grows excessive, since it would produce an unobserved outwards acceleration of the gas.

(3) Weakness of gamma radiation. As B is imagined to be smaller, the number of electrons present must be increased, and their energies increased also. The fast electrons undergo Compton scattering with the synchrotron photons present in the nebula, yielding high-energy gamma rays (Gould, 1965). We will discuss this in more detail later. Present predictions of the flux from this process are only a factor 30 below the upper limits set by gamma-ray observations of the Crab Nebula. Therefore the number of electrons at each energy cannot be raised by more than a factor of 30 above the minimal number required to explain the synchrotron flux. Thus the field cannot be reduced more than a factor of ten.

With all these considerations in mind, the average field in a uniform model of the Crab Nebula must be set within one order of magnitude of the field that minimizes the total energy and equates the energy of the present electrons with that of the field, namely 5×10^{-4} gauss. Similar considerations apply to other supernovae (May, 1968) and to galactic sources such as Virgo A, but with slightly different quantitative results (Felten, 1968).

In the case of Taurus A there is another clue, not present for most other sources. The synchrotron power spectrum has a spectral index of -0.26 from low radio frequencies up to the infrared, and then increases to about -1.0 for frequencies from the visible up to gamma-rays of at least 400 kilovolts. The bend occurring at $\nu \cong 2 \times 10^{13}$ sec^{-1} may be accounted for as follows. Electrons responsible for synchrotron radiation below this frequency have a lifetime for energy loss exceeding the age of the nebula, so all the electrons produced in the history of the nebula are still present. Those of higher energy, however, have shorter lifetimes in proportion to

3*

E^{-1}; so only those produced in recent times, within an interval of length $\Delta t \propto E^{-1}$ extending back from the present, have been able to retain their energy. Therefore, the electron spectrum steepens beyond a certain energy that depends on the field strength B. From this, B can be calculated. Going through this procedure will serve to introduce a few relations that are useful to subsequent discussions.

The average rate of energy loss of an electron moving at random pitch angles in a field of strength B is

$$\frac{dE}{dt} = -bE^2 \quad \text{with} \quad b = \frac{4}{9} \frac{r_0^2 c B^2}{(mc^2)^2} \quad (r_0 = e^2/mc^2).$$

Therefore after a time t, particles that were injected with energy E_0 have a residual energy related to E_0 by $E_0 = E(1 - Ebt)^{-1}$ provided $Ebt < 1$. *No* particles can be left with energy exceeding $(bt)^{-1}$. If the age of the nebula is T_0 and particles have been injected constantly with a spectrum $E_0^{-\beta} \, dE_0$, the number remaining at energy E should be

$$N(E)\, dE = E^{-\beta} dE \left[\frac{1 - (1 - bT_0E)^{\beta - 1}}{(\beta - 1)\, bE} \right] \quad \text{provided} \quad bT_0E < 1 \qquad \text{(a)}$$

$$\cong T_0 E^{-\beta} \, dE \quad \text{provided} \quad bT_0E \ll 1 \qquad \text{(b)}$$

$$= \frac{1}{b(\beta - 1)} E^{-(\beta + 1)} \, dE \quad \text{provided} \quad bT_0E > 1. \qquad \text{(c)}$$

If, in addition, there was a dowry supplied at $t = 0$ with a spectrum $= aE_0^{-\beta_0} \, dE_0$, the following residue should be added to expressions (a) and (b):

$$N_0(E)\, dE = aE^{-\beta_0} \, dE(1 - bT_0E)^{\beta_0 - 2},$$

which is approximately $aE^{-\beta_0} \, dE$ when $bT_0E \ll 1$. It is more realistic to consider an injection rate that decreases with time. The effect of doing so is to steepen the spectrum at the high energies without having much effect below the energy where the bend appears, and without affecting the apparent position of the bend.‡ The bend is located by fitting straight lines (in logarithmic plots) to the low-energy and high-energy expressions, (b) and (c).

---

‡ Without the supplementary spectrum injected at $t = 0$ or an injection rate that changes with time, one sees from expressions (*b*) and (*c*) that the electron spectral index would have to change by 1.0. The index of the synchrotron power spectrum would then have to change by 0.5. In fact, the change is more nearly 0.75; thus the electron number spectrum must steepen by 1.5 in the exponent.

The intersection of such lines identifies an energy E_c, and corresponding Lorentz factor γ_c:

$$\gamma_c = [bT_0mc^2(\beta - 1)]^{-1} = 0.05B^{-2}$$

(using $T_0 \cong 3 \times 10^{10}$ sec and $\beta = 1.52$ for Taurus A). The frequency ν_1 at which the bend occurs, is related to γ_c by

$$\nu_1 = \frac{0.34Be}{2\pi mc}\gamma_c{}^2.$$

Inserting $0.05B^{-2}$ for γ_c and 2×10^{13} sec^{-1} for ν_1 leads to

$$B = (0.05)^{2/3}\left(\frac{0.34e}{2\pi mc\nu_1}\right)^{1/3} \simeq 5 \times 10^{-4} \text{ gauss (for Tau A)},$$

in close agreement with estimates made above on other bases.

Lumpy Fields and Electron Lifetime

Even the extension of the synchrotron spectrum to frequencies as high as the ultraviolet ($\sim 10^{15}$ cycles/sec) has caused concern because of the shortness of electron lifetime for energy loss:

$$T = E/(-dE/dt) = (bE)^{-1} = \left[\frac{4}{9}\frac{r_0^2 cB^2E}{(mc^2)^2}\right]^{-1}$$

$$\simeq \frac{9}{4}\left(\frac{0.34e}{2\pi mcv}\right)^{1/2}\frac{mc^2}{r_0^2 cB^{3/2}}.$$

For $B = 1$ milligauss and $\nu = 10^{15}$ sec^{-1} (4.1 eV quantum energy), $T = 7.6 \times 10^8$ sec $= 24$ years, much less than the duration of the nebula; therefore contemporary replenishment of the electron energy is essential, as well as acceleration throughout the nebula.

The situation grew more critical when x-rays were discovered that appeared to represent a continuation of the optical synchrotron radiation, and the x-ray spectrum was found to extend as an uninterrupted power law to at least 400 keV. Since $T \propto \nu^{-1/2}$ and 400 keV is 10^5 times the quantum energy considered above, we find in this case $T = 2.5 \times 10^6$ sec $= 4$ weeks. While the electron energy needed to produce the UV was 5×10^{11} eV, for these x-rays it is 1.6×10^{14} eV in a one-milligauss field. The cyclotron radius for such electrons is 5×10^{14} cm and it takes 1.3 days to complete one circuit. The electrons therefore do not have time to propagate more than about 10^{-3} of the diameter of the nebula from the place or places where they are accelerated.

This quandary turned from worrisome to critical upon discovery of a pulsing neutron star in each of two supernovae (the Crab and the SN in Vela), and the discovery that the pulsar in the Crab Nebula was emitting not only radio signals but also pulsed visible light and x-rays. Measurement of the slowing-down rate, furthermore, showed that the loss of rotational energy by the neutron star closely matches the total radiant power output of the nebula. Therefore it became virtually certain that the source of energy is the neutron star, and that the electron acceleration occurs in a very small volume (relative to the nebula) around the neutron star. But it is known that the unmodulated x-rays are emitted from a region of large radius, about 10^{18} cm. How, then, do electrons with such short life-time manage to get so far from the source of energy?

The acceleration mechanism proposed earlier by Colgate and White (1966), in which the expanding shell of a supernova attains relativistic speed, did not solve the problem, because it did not provide a *constant* source of high-energy electrons at large distances from the center.

Since the detailed mechanism of energy transfer from the neutron star to electrons and protons is not yet established, the problem may have more than one possible solution. However, it has been suggested many times that one can prolong the lifetime of high-energy electrons in the nebula by postulating that strong magnetic fields exist in numerous concentrated regions, very small compared with the nebula. The electrons lose energy rapidly in these regions, but may propagate through the nebula for a long time before entering one: partly because of the small cross section of the knots and partly because of magnetic reflection at the boundaries. In the remainder of the nebula, the field is imagined weak enough to allow the electrons to diffuse through the system before losing their energy.

There are two serious obstacles to this solution. First, if one pro-longs the average lifetime for energy loss by the high-energy particles, many more of them must be stored in the nebula, where by Compton scattering they would produce more high-energy gamma rays than are observed. Secondly, if most of the emission occurs in regions of strong field, the radio waves should exhibit circular polarization As discussed above, observational limits on this polarization apparently require the average field in the emitting regions to be less than 10^{-3} gauss: i.e., insignificantly more than in a uni-form-field model.

Total Energy

In addition to the uncertainty in the detailed distribution of par-ticles and fields in an object like the Crab Nebula, there is also some uncertainty in its distance. Currently 1700 pc (5.2 × 10²¹ cm) is favored, though a short while ago 1100 pc (3.4 × 10²¹ cm) was thought right. At 1700 pc, the minimum energy present in particles and magnetic field is 10^{48} ergs in each, while the radiated power is 5×10^{37} ergs/sec. At this rate, over the age of the nebula the radiated energy has been 1.5×10^{48} ergs, about the same as the remaining electron energy.

However, there is reason to believe the radiated power has been decreasing with age of the nebula. The agreement between the present radiated power and the rate of loss of rotational energy of the pulsar, and the agreement between the latter and the ratio of its residual energy to its age, suggests that $dE/dt \simeq -E/t \propto t^{-2}$. The slower rates of deceleration of the pulsars that have longer periods are roughly consistent with this. The total energy that has been emitted is then 10^{51}–10^{52} ergs, on the order of the initial energy of rotation. Other models of supernovae also predict a diminishing radiant power owing to reduction of the field and total particle energy as the nebula expands.

It is most urgent to detect x-rays from a supernova in its beginning phase: both the burst associated with the exploding shell and the subsequent radiation during the first year or so of its existence. Doing this for SN occurring in external galaxies is within existing capability. Without such observations our present estimates are speculative. The total electron energy that has been dissipated in radiation in the Crab Nebula, for instance, is certainly large compared with 10^{48} ergs, but it may lie anywhere between 10^{49} and 10^{52} ergs.

There are other ways, as well, of reducing the uncertainties about the electrons and fields in the nebula, and these investigations are in progress:

(1) Fine-resolution radio profiles, related to the spatial distribution of magnetic field and electrons in the nebula.

(2) Attempts to measure gamma rays arising from inverse Compton scattering of the electrons. The *radiation* field within the nebula is rather well known from the radiation arriving at the Earth. Therefore the inverse Compton gamma-ray yield will be a direct measure of the electron spectrum in the nebula, independent of assumptions about the strength of the magnetic field. Then the synchrotron spectrum will permit the magnetic field to be deduced.

Heavy Particles

We turn now to inferences about the high-energy protons and other nuclei produced in a supernova. As mentioned before, knowledge of these is severely limited. However, indications are being sought by high-energy gamma-ray experiments. The energetic nuclei, in collisions with atoms of the background gas, produce neutral pions that decay into gammas that can be detected at the Earth. Gamma rays of energy greater than about 10^8 eV have been sought with counter and spark-chamber experiments that have set an upper limit of about 3×10^{-5} cm^{-2} sec^{-1} to the flux at the Earth, or 10^{40} gammas per second emitted at the Crab Nebula (see, e.g., Fazio *et al.*, 1968); and detectors of Čerenkov light from small air showers have been applied to search for gammas above 10^{12} eV (e.g., Fegan *et al.*, 1968) with a tentative positive result of 10^{-10} photons cm^{-2} sec^{-1} (3×10^{34} per second at the Crab Nebula). To be conservative, the latter result must be treated as an upper limit. Similar upper limits to the flux reaching the Earth at 10^8 and 10^{12} eV have also been established for other sources.

Assuming an average gas density of 10 nuclei/cm^3 in the Crab Nebula and a cross section of 30 millibarns, the mean reaction time per fast proton is 10^{14} seconds. Hence the above data set limits of 10^{54} and 3×10^{48}, respectively, on the number of protons in the Crab Nebula with energy above 10^9 and 10^{13} eV. There is promise that these limits will be reduced in the near future, since detectors of greater sensitivity in both energy ranges are under construction. Hopefully they will yield positive values of the flux instead of just limits. In the energy range 10^7–10^8 eV, it is possible to distinguish gammas arising from pion decay from those made by electrons through Compton scattering, because the pion mechanism always generates a characteristic peak in the spectrum at 68 MeV (half the neutral-pion rest energy).

Meanwhile, estimates of the number of heavy particles can be made, based on analogies and on the electron spectrum of which we have direct knowledge as discussed above.

In the Earth's magnetosphere, the total kinetic energy of the protons is ~ 100 times that of the electrons. Among solar energetic particles (ESP) also, the energy of all the heavy particles is much greater than that of electrons. Among galactic cosmic rays (which probably originate mainly in supernovae), the number of nuclei is ~ 100 times the number of electrons of the same energy (indeed, the number spectra are remarkably parallel, two decades apart in intensity). By analogy, since the energy of the electrons now in the Crab Nebula is 10^{48} ergs or more, one is led to suggest that the heavy particles were probably given at least 10^{50} ergs.

Incidentally, the kinetic energy of the expanding gas of the Crab Nebula is now about 10^{49} ergs; the original optical flare released about 10^{50} ergs in radiation; and the total energy of the explosion has been estimated as 10^{52} ergs. If supernovae occur once every 100 years in the Galaxy, and supply the principal part of the total energy of galactic cosmic rays (requiring a total energy input of 10^{41} ergs per second), the heavy particles must receive, on the average, 3×10^{50} ergs in each event.

The kinetic energy of the neutron star formed in the collapse associated with the supernova explosion was about 10^{52} ergs if the angular momentum before collapse was similar to that of the Sun. It now appears that the rotational energy is dissipated largely by efficient generation of relativistic particles. One is thus led to expect 10^{51} ergs or more to be given to these particles.

All of these bases of estimation lead to remarkably consistent results in the neighborhood of 10^{50}–10^{51} ergs of total particle energy. The electrons may even have received an energy similar to that of the heavy particles, but radiated most of it away.

A remarkable feature of explosions on an astronomical scale, that always appears in these estimates. is the extraordinary efficiency with which the energy is converted to high-energy particles: 10^{34} grams of matter conspire in such a way as to give several percent of the energy to one *millionth* of all the particles—as in the coherent socio-economic processes that conspire to make the rich richer and the poor poorer in society.

The consistency of all these estimates is marred by one prediction. If the proton injection spectrum in the Crab Nebula (and other similar objects) is parallel to the electron spectrum but the protons are 100 times more numerous on injection and the spectrum is not steepened by radiative losses, one has a basis for calculating the expected gamma-ray yield, resulting from pion production in collisions with the gas. The result of such calculations is well below the observational limits at 10^8 eV, but not so at 10^{12} eV: the prediction is an order of magnitude above the limits set by the Čerenkov light experiments.

One proposed way out of the dilemma was the lumpy-field model, which allowed one to suppose (since it permitted stronger magnetic fields) that the electron spectrum did not extend as high as 10^{12} eV, and the proton spectrum might also terminate near this energy. To me this suggestion is implausible, especially since the discovery of very high energy x-rays that seem to be an extension of the synchrotron spectrum. Without a lumpy field, electron energies must extend beyond 10^{14} eV, and one can only reduce them in proportion to $B^{-1/2}$. An attempt to go far in this direction puts excessive demands on the total energy of the particles stored in the system: indeed, it greatly *raises* the number of lower-energy particles that are required to be there, and with it the estimate of gamma-ray yield at the 10^8-eV level.

A better way out exists. It has been pointed out that the field energy in the Crab Nebula is only 10^{48} ergs while the proton energy is estimated to be 10^{50} ergs. The field lines, of course, are rooted in the gas, but there is not enough gas to contain the pressure of all those protons if, in fact, they are trapped in the nebula: they would have caused a large radial acceleration which has not, in fact, occurred. Therefore the field structure must be open enough to have permitted more than 90% of the protons to have escaped into the surrounding interstellar space. There the gas density is lower by an order of magnitude and so the expected gamma-ray yield is sufficiently reduced to be below the present limits of detectability.

The latter explanation seems very plausible. However, the protons have not had time for most of them to diffuse more than 50 light years or so from the nebula. Therefore, with a little more detector sensitivity one might expect to see a *diffuse* source of high-energy gamma rays, about one degree in diameter, surrounding the Crab Nebula. It will be interesting to see if this is observed in the near future.

3.2.   Radio Galaxies and Quasars: Cutoff due to Recession

We will not attempt to focus discussion as closely on radio galaxies as we have on SN, but will merely point out a few similarities and contrasts.

First, the transmission of particles to Earth from radio galaxies not only is much less efficient than from SN, but probably involves much more time delay. It is believed that the active phase of galactic evolution lasts only 10^6–10^8 years. If so, particles are only reaching us from external galaxies that are no longer visibly explosive.

The scale of radio galaxies and the energies involved, of course, are much larger than corresponding quantities for the SN. The radio power may be as large as 10^{45} ergs/sec (as for Cygnus A), in contrast to 10^{36} erg/sec from the Crab Nebula. The total radiative output from such a galaxy in its active life exceeds 10^{60} ergs. If the x-ray power greatly exceeds the radio power as it appears, the total luminosity may exceed 10^{46} ergs/sec, and the energy output exceed 10^{61} ergs.

The inference of high-energy particles in these tremendous infernos is made by the same arguments and types of evidence that were applied to SN.

About quasars the major controversy, as to whether the big redshifts are reliable indicators of cosmological distance, still rages. The majority vote is for the cosmological interpretation. If so, they too, although much smaller than radio galaxies in size, have luminosities on the order of 10^{46} ergs/sec (assuming that focussing by gravitational lenses is not what accounts for their apparent luminosity). Quasars are distinguished by the characteristic blue continuum radiation which is indicative of synchrotron radiation. Therefore a large part of their energy is in high-energy particles.

Rapidly receding objects at cosmological distances can be (and are) sources of electromagnetic radiation reaching the Earth—neutrinos also. But not of charged particles that reach the Earth. The average velocity of charged particles along a straight line is significantly less than c because of the scattering by magnetic fields. How *much* less than c the velocity is in intergalactic space depends on the particle energy, the intergalactic field strength, and the distance or time over which the average is taken. Even for particles of very high energy (say 10^{20} eV), the average diffusion velocity, taken over a time of 10^8 years, is probably very small compared with c. When the recession velocity of the source exceeds the mean diffusion velocity of the particles in the medium surrounding the source, none of the particles diffuse *towards* the Earth: they all recede, along with the region of the Universe in which they are embedded.

Therefore no particles except neutrinos (and photons) are likely to communicate with us from beyond the local supercluster. We can only see the charged particles from afar by the electromagnetic radiation with which they signal to us.

4. INTERACTIONS IN THE INTERSTELLAR MEDIUM

By many well-known processes, the interstellar medium acts as a source of high-energy particles and photons. In the introductory section, we have already discussed how *acceleration* can occur by reflection of particles from advancing inhomogeneities in the magnetic field, transported by gas clouds. Also contributing to acceleration is a betatron process: the Lorentz force due to local time variations of the field, induced for instance by hydromagnetic waves.

In addition to gradual acceleration, however, there are numerous reactions by which energy is given abruptly to some form of secondary cosmic radiation, and that is what will be discussed in the present section. These reactions occur between fast particles and targets of three types, as follows.

A. Collisions with the gas

(1) The "knockon" process—elastic interactions of charged particles with atomic or free electrons, whereby the latter are given kinetic energy. This is a dominant process in the origin of electrons of moderate energy, up to about 15 MeV.

(2) Negatron and positron emission in beta decay of neutrons and unstable nuclei, produced in strong nuclear reactions.

Although the motion of the parent nucleus makes it possible for the electrons and positrons to have high energy in the laboratory, they can never have more than a very small fraction of the total energy of the parent nucleus. This process contributes dominantly only to the positron component, and only below a few MeV.

(3) Electron and positron generation by pion and kaon production and decay.

K production has a higher threshold than π production, and is about six times less frequent even at very high energies. The most frequent reactions leading to electrons or positrons are therefore the sequence

$$p + p \rightarrow p + n + \pi^+,$$

$$n \rightarrow p + \bar{\nu}_e + e^- \text{ (low-energy electrons)},$$

$$\pi^+ \rightarrow \mu^+ + \nu_\mu,$$

$$\mu^+ \rightarrow e^+ + \bar{\nu}_\mu + \nu_e.$$

Near threshold, this yields only positrons, not electrons (except for the inefficient neutron channel, which gives very little energy to the electron). On the average, the positron gets one-fourth of the total energy of the positive pion.

Because the initial p–p interaction is a strong one and involves the most abundant particles of both the interstellar medium and the cosmic rays, and also because the reaction is efficient in transferring high energy to the secondaries (in contrast to the knockon and beta decay processes), the pion decay route dominates heavily over mechanisms (1) and (2) in production of high-energy positrons. However, it yields very few positrons below five MeV.

Collisions involving heavy nuclei of the interstellar gas, or heavy nuclei of cosmic rays, or both, can yield negative electrons, essentially by

$$n + p \rightarrow p + p + \pi^- \quad \text{and} \quad n + n \rightarrow n + p + \pi^-.$$

However, unless both the cosmic-ray nucleus and gas nucleus are heavy, there is still a greater yield of positrons than electrons, because *np* reactions yield π^+ and π^- equally, but half of the reactions of a proton with a heavy nucleus are with other protons, and yield predominantly positive pions.

When the incident particle has an energy exceeding one GeV, *multiple* pion production begins to be important. Then the production of electrons is no longer overwhelmed by positron production, though the two never attain equality. Also, kaon production begins at about this energy, and leads to e^+ and e^- by various decay modes; but at energies near threshold, K^+ production is much more frequent than K^-, and this imbalance continues to accentuate the production of positrons.

In the atmosphere, the relative number of positive and negative muons reaching sea level is 5 to 4 with remarkable independence of energy up to at least 500 GeV. The energy required to penetrate the atmosphere sets a lower limit of 1.5 GeV on the muon energies for which this observation applies. These particles are produced in interactions of the primary cosmic rays and secondary hadrons with atmospheric nuclei containing equal numbers of neutrons and protons. Taking into account the proportions of heavy nuclei among the primaries, and assuming charge symmetry both in *p–n* collisions and in the nuclear cascade following the first interaction, one can deduce that in *p–p* interactions leading to positrons and electrons between 0.5 and 200 GeV, 3/4 of the secondaries are positive. It then follows—taking into account the composition of both the cosmic rays and the interstellar gas—that the positron to electron ratio among the secondaries in the interstellar medium is 2 to 1. At energies below 0.5 GeV the positive excess is even greater, until low enough energies are reached (below about 30 MeV) for knockon electrons to become appreciable.

(4) Gamma-ray and neutrino production.

Concomitantly with the production of electrons and positrons by pion and kaon production and decay, there is a production of gamma rays by the decay of neutral pions. Half as many neutral pions as charged ones are produced, but each π^0 creates *two* photons, so as many photons are generated as e^+ and e^- together. However, electrons, on the average, receive 1/4 of the pion energy while the photons receive one-half. Therefore, since the spectral index of the cosmic rays is about 2.6, the number of secondary photons at a given energy is $2^{1.6} \simeq 3$ times the combined number of electrons and positrons.

The decay reactions leading to muons, electrons, and positrons also produce four kinds of neutrinos: a ν_μ–$\bar{\nu}_\mu$ pair for each charged pion, a ν_e for each positron, and a $\bar{\nu}_e$ for each electron. At energies large compared with 100 MeV, the spectra of these neutrinos are similar to those of the electrons and positrons.

(5) Bremsstrahlung.

Gamma rays are also produced by bremsstrahlung of fast electrons scattered by nuclei of the gas. We will discuss this more fully, as well as gamma production by π^0 decay, in a later section.

(6) Rare nuclei produced by spallation.

As mentioned in the first section, nuclei that are rare in the primary cosmic-ray sources are produced by fragmentation of heavier-cosmic-ray nuclei in collisions with the interstellar gas. Deuterium, He^3, the L-nuclei (Li, Be, B), and other rare types (such as fluorine) are produced in this way.

(7) Antiprotons.

The threshold for antiproton production in p–p collisions is about 6 GeV. At energies four or five times this high, the production of antiprotons is about 10^{-2} times that of pions, and is no longer changing rapidly with energy. It will appear, below, that the intensity of electrons and positrons originating in the interstellar medium is about 10^{-3} of the intensity of protons of the same energy. We may therefore anticipate a flux of antiprotons on the order of 10^{-5} times the proton flux, even if no antiprotons come from primary sources. No experimental evidence yet contradicts this expectation.

B. Processes involving photons as target or beam particles

(1) Elastic scattering, $e + \gamma \to e + \gamma$.

The elastic scattering of photons by free electrons is known as Thomson scattering as long as the condition is met that the photon energy in the frame of reference of the electron is very small compared with $m_e c^2$. The cross section is then a constant, $(8/3)\,\pi r_0^2 = 6.64 \times 10^{-25}$ cm^2, where $r_0 = e^2/mc^2$ is the classical electron radius; and the angular distribution of the scattered photons with respect to their initial direction is $1 + \cos^2 \theta$ in the frame of reference of the electron. Protons and other nuclei scatter photons similarly, but since the cross section varies as the inverse square of the mass, the scattering by heavy particles is of negligible importance.

The exact formula for the energy of the scattered photon, W', in the original reference frame of the electron is

$$W' = \frac{W}{1 + (1 - \cos\theta)\,W/mc^2},$$

where W is the photon energy prior to scattering. $W'_{max} = W$ and $W'_{min} = W/(1 + 2W/mc^2)$. Under the conditions for Thomson scattering, there is very little change in photon energy, only a change in direction.

When W is not small compared with mc^2, the cross section is smaller, the angular distribution is more peaked in the forward direction, and the quantum energy is substantially reduced in the original frame of reference of the electron. The process is then referred to as Compton scattering. The accurate cross section is given by the Klein–Nishina formula:

$$d\sigma = \pi r_0^2 \frac{mc^2}{W} \frac{dW'}{W'} \left[1 + \left(\frac{W'}{W} \right)^2 - \frac{W'}{W} \sin^2 \theta \right].$$

The total cross section can be approximated by

$$\sigma \simeq \pi r_0^2 \frac{mc^2}{W} \ln \left(1 + \frac{8}{3} \frac{W}{mc^2} \right),$$

which is accurate asymptotically for large and small values of W/mc^2, and high by about 13% in the intermediate region.

The cross section is not high enough, in view of the interstellar electron density, to create an appreciable interstellar opacity for photons of any energy in the present epoch. However, the *photon* density in space is high enough—especially that of microwave and infrared photons—so that electrons which diffuse for millions of years undergo scattering very many times. When the original photon energy in the laboratory system is low and that of the electron is high, the consequence is to reduce the electron energy seriously, and to provide a source of energetic x and gamma rays. The process under these conditions is referred to as *inverse Compton effect*.

Within intense sources of synchrotron or other electromagnetic radiation, the photon density is often much higher than in interstellar space. If energetic electrons are present, as they are in the nonthermal sources, the inverse Compton effect provides a mechanism for production of much more energetic photons than those which the same electrons can produce by magnetic bremsstrahlung.

We will be concerned in this discussion only with highly relativistic electrons, for which the Lorentz factor, $\gamma = E/mc^2$, is very large compared with 1. When an electron of velocity βc encounters an isotropic flux of photons of energy W_0, the average energy of these photons in the electron frame of reference is $\bar{W} = \gamma W_0 (1 + \beta^2/3) \simeq (4/3) \gamma W_0$. The average direction, θ_0, of the photons in this frame, with respect to the line of relative motion of the frames of reference, is given by $\langle \cos \theta_0 \rangle = \beta$, hence $\bar{\theta}_0 \simeq \gamma^{-1}$: i.e., the collisions are essentially head-on. After scattering through angle θ in the electron frame of reference, the photon energy in the laboratory is therefore

$$W_2 = \gamma W'(1 - \beta \cos \theta) \simeq \frac{4}{3} \gamma^2 W_0 (1 - \cos \theta) \left[1 + \frac{4}{3} \frac{\gamma W_0}{mc^2} (1 - \cos \theta) \right]^{-1}.$$

Consider interactions with a radiation field such as the 2.7° background radiation, for which $\bar{W}_0 = 2.7kT = 6.3 \times 10^{-4}$ eV and the photon number density is 400 cm^{-3} (energy density 0.25 eV/cm^3). The second term in the bracket does not become serious until $\gamma \gtrsim 6 \times 10^8$; i.e., until the electron energy approaches or exceeds 3×10^{14} eV. When this term is negligible, the conditions for Thomson scattering prevail: the angular distribution is symmetric around $\theta = 90°$, and $\bar{W}_2 \simeq (4/3)\gamma^2 W_0$. In interactions with longer radio waves, this result is valid to even higher electron energy, and hence to *higher secondary photon energy* as well, whereas in interactions with visible light, the Thomson conditions prevail only to electron (and secondary photon) energies of about 60 GeV. Even in the latter case, there is a wide range of validity of this approximation.

For highly relativistic electrons, the factor $(4/3)\gamma^2$ is ordinarily very big. For instance, electrons of one GeV scatter the 2.7° background radiation as x-rays of about 3 keV (4 angstroms), while electrons of one TeV (which may be responsible for the blue synchrotron light in the Crab Nebula) scatter this microwave radiation as gamma rays of 3 GeV.

Because of the proportionality between the scattered photon energy and the *square* of the electron energy, an electron spectrum going as $E^{-b}\,dE$ produces a spectrum of scattered photons distributed in number as $W^{-\alpha}\,dW$, where $\alpha = (b + 1)/2$. This continues up to an energy W_{max}, determined either by the upper limit of the electron spectrum ($W_{max} \simeq \gamma^2_{max} W_0$) or by the breakdown of the Thomson approximation ($\gamma W_0 \simeq mc^2$). Beyond the latter limit, the secondary energy only increases with the first power of the electron energy ($W_2 \simeq E$) and the cross section diminishes as $E^{-1} \log E$; hence the spectral index, α, is approximately doubled.

The generation of secondary radiation is accompanied by energy loss of the electrons. The mean free path is $(n\sigma)^{-1} \simeq (8\pi r_0^2 n/3)^{-1}$, which, for instance, is 4000 light-years for scattering by the 2.7° microwaves, while the mean energy loss per scattering is $(4/3)\gamma^2 W_0$, hence the rate of energy loss is

$$\frac{dE}{dt} \simeq -\frac{32}{9}\pi r_0^2 \frac{E^2}{(mc^2)^2} cu,$$

where u is the energy density of the background radiation. Expressing u in eV (cm)$^{-3}$ and E in GeV, this relation becomes

$$\frac{dE}{dt} = 1.02 \times 10^{-16} uE^2 \text{ per second}.$$

The time interval for loss of half the electron energy is then $0.98 \times 10^{16}(uE)^{-1}$ seconds, or $3.1 \times 10^8(uE)^{-1}$ years. For instance, with

$u = 0.25$ eV/cm^3, this becomes 2.5 million years (approximate diffusion time in the disk of the Galaxy) at $E = 500$ GeV, and 10^{10} years (age of the Universe) at 0.12 GeV. The addition of other components of background radiation and other kinds of energy loss, of course, shortens this absorption time, or reduces the initial energy corresponding to a given time scale.

(2) Photodisintegration.

Heavy nuclei undergo γ–n reactions, followed by beta decay of the neutron, and positron emission by the nucleus, resulting in a gradual breakup of the nucleus into separate protons, if the nucleus is immersed in a flux of gamma rays above the neutron binding energy (about 8 MeV). The cross sections are of order 10^{-25}–10^{-26} cm^2, depending on the atomic weight of the nucleus. In order that the time scale be short enough to result in significant change of the nuclear composition of cosmic rays, the photon density must be high; hence the process must make use of visible and infrared light or microwaves (number density 10^{-1} to 10^3 cm^{-3}). This is indeed possible provided the nuclei have sufficiently high Lorentz factor, such that the photons have energy exceeding 8 MeV in the frame of reference of the nucleus. Thus the Lorentz factor must be 10^7–10^{10}, or the energy per nucleon 10^{16} (for light) to 10^{19} eV (for interaction with the microwaves). A consequence can be the depletion of heavy nuclei among the very high energy cosmic rays. The effect has been evaluated quantitatively by F. W. Stecker (1969).

(3) Photoproduction of pions.

There are not enough high-energy gamma rays in interstellar space for photoproduction against ambient protons of the gas to be an important secondary production mechanism, nor are there enough ambient protons to create significant opacity of the Universe to high-energy photons. But the density of *low-energy* photons is great enough to make the process consequential when they can participate in it. This happens for the 2.7° background, for instance, when it encounters nuclei of energy exceeding 10^{20} eV per nucleon. In the frame of reference of a proton having such an energy, these photons are gamma rays a little above threshold for production of $n + \pi^+$ or $p + \pi^0$. The cross section is several hundred microbarns, making the mean free path about 10^{25} cm, short compared with the wandering path length (about 10^{28} cm) travelled by a proton if there are no absorption processes. (Protons of such high energy are not confined in the Galaxy, and are assumed to spend most of their life in intergalactic space, where the only other important loss is by the cosmological redshift.)

Attention was first called to the effect on the high-energy spectrum by Greisen (1966) and by Zatsepin and Kuzmin (1966); it has been discussed subsequently in more detail by Greisen (1967) and Hillas (1968), and evaluated with more accurate cross sections by Stecker (1968).

There are two effects of the interactions: the production of the pion decay products (neutrinos, electrons, and photons) at energies on the order of 10^{19} eV, and the reduction of energy of primary nuclei that have attained 10^{20} eV per nucleon or more. Each reaction only removes on the order of 20% of the nucleon's energy; hence the energy absorption length is about five times the interaction length. The absorption time, according to Stecker (1968), is about 10^{16} sec for protons of 10^{20} eV, and reaches a minimum of 10^{15} sec for protons of 10^{21} eV. However, if in addition to the 2.7° background, the Universe contains a lot of infrared radiation (Shivanandan, Houck, and Harwit, 1968), the threshold energy may be lower, and the absorption time may be substantially shorter for the particles that are above the threshold. In the remote past, when the temperature of the background radiation was higher, both the threshold energy and the absorption time were lower.

(4) Pair production: photons against nuclei [Greisen (1966, 1967), Hillas (1968), Encrenaz and Partridge (1969)].

As in the above examples of photodisintegration and pion photoproduction, the production of electron–positron pairs is only frequent enough to be significant when low-energy photons can participate. This is possible when a high-energy nucleus has a Lorentz factor exceeding the ratio of the electron rest-energy, mc^2, to the laboratory energy of the photon, W_0. The 2.7° radiation can interact with protons to produce pairs, for instance, when the proton energy exceeds 10^{18} eV. The energy of the electron–positron pair is then about 10^{15} eV, and this energy is lost by the proton. Although the fractional energy loss is small, the frequency of the events is moderately large: the cross section is on the order of Z^2 millibarns, where Z is the atomic number of the nucleus. Thus for protons the mean free path is about a million light-years, and the energy absorption time is about equal to the age of the Universe.

For primary particles of cosmological age, the absorption due to this process is enhanced by the increased number density and quantum energy of the thermal photons in the remote past.

The positrons and electrons that are created do not last long, because of the inverse Compton effect that has been discussed above: their mean time for absorption is only about 10^4 years. Hence the equilibrium number of 10^{15}-eV electrons is extremely small. But the Compton process yields photons of energy exceeding 10^{14} eV. The fate of these quanta is discussed in the next paragraph.

(5) Pair production: photons against photons.

The process, $\gamma + \gamma \to e^+ + e^-$, is the inverse of electron–positron annihilation and can occur whenever the product of the photon energies exceeds $(mc^2)^2$. The cross section has a maximum of 1.7×10^{-25} cm^2 at an energy about twice the threshold value, and remains within a factor 3 of the maximum from an energy very close to threshold to one 15 times the threshold value.

Thus, starlight of energy around one eV can interact with photons above 260 GeV to make electron–positron pairs. The resulting opacity is greatest around 10^{12} eV. In the disk of the Galaxy the average optical photon density is about 0.3 per cm^3, hence the mean free path is 3×10^{25} cm, much greater than the galactic diameter, and the opacity is negligible. In intergalactic space the optical photon density is about 10^{-2} per cm^3 and hence the mean free path is 10^{27} cm. Since this is ten times less than the Hubble length, the Universe is opaque at 10^{12} eV. Abundant infrared photons from relatively cool stars and galaxies extend this opacity to at least 10^{13} eV.

The very high number density (400 cm^{-3}) of the 2.7° background makes the Universe extremely opaque to photons above the threshold for pair creation against these microwaves. The minimum mean free path is about 2×10^{22} cm, smaller than the Galaxy, at about 10^{15} eV. Because of the width of the Planck distribution, significant opacity of the Universe sets in rapidly above 10^{14} eV; and the low-frequency tail of the thermal distribution combines with the slow decrease of the cross section with energy to maintain a high opacity up to energies beyond 10^{19} eV. With increasing energy of the high-energy photon participant, more and more of the radio-frequency background in the Universe becomes capable of initiating the pair creation. Thus, there is no energy above 10^{14} eV for which the Universe is transparent to photons.

There may be an electromagnetic "window" between 10^{13} and 10^{14} eV, where the optical photons do not provide great opacity; but even this part of the spectrum may be strongly absorbed if the infrared detected by Shivanandan et al. (1968) is universal. In any case, below about 3×10^{11} eV (down to the soft x-ray region), the Universe is transparent. Detection of primary photons in the energy range 10^{13}–10^{14} eV may be one of the most powerful ways of probing the infrared background flux in remote parts of the Universe.

The great opacity of the Universe for high-energy photons, owing to the thermal background radiation, was first noted by Jelley (1966) and by Gould and Schréder (1966). The theory had previously been presented in detail by Nikishov (1961–62), and a detailed analysis, taking into account the effects of all parts of the electromagnetic spectrum, has been published by Gould and Schréder (1967).

At the end of the previous section, it was noted that high-energy protons (above 10^{18} eV) upon interacting with the 2.7° background, yield electrons and positrons of about 10^{15} eV, which are rapidly absorbed by inverse Compton scattering off the 2.7° thermal photons. The recoil photons so produced are mainly between 10^{14} and 10^{15} eV. According to the present section, these photons continue the showering process by generating further electron–positron pairs, again by interacting with the thermal background photons.

Thus, the universal 2.7° background radiation serves as an interacting medium in which cascade showers can occur. The reciprocal processes are inverse Compton scattering and photon–photon annihilation. The "radiation length" is about 2×10^{22} cm. The multiplication continues until the photons are under 10^{14} eV in energy. The electrons then continue to lose energy, making more recoil photons; but reciprocation by pair production has practically ceased, unless there is an infrared background component that is much more abundant than the optical radiation. Barring this eventuality, about half of the total energy of all electrons and photons generated above 10^{14} eV is deposited and stored up in the Universe in the form of photons of energy between 10^{13} and 10^{14} eV. One could seek evidence of this cascade process by looking for characteristic changes in slope of the diffuse primary photon spectrum. A steepening should begin above 300 GeV, due to opacity created by optical photons. This should be followed by a pronounced flattening of the spectrum between 10^{13} and 10^{14} eV, while beyond 10^{14} eV the spectrum should become extremely steep, and only a highly anisotropic flux of galactic origin should remain beyond about 2×10^{14} eV.

C. Interactions with interstellar fields

The interaction with magnetic fields is mentioned in the present section primarily for the sake of completeness, since the only important process is magnetic bremsstrahlung, which has already been discussed in Sec. 3. It is worth while at this juncture, however, to note the points of similarity between magnetic bremsstrahlung and inverse Compton scattering, which has been reviewed in the present section.

As long as the electron energy is not too high, and the directions of motion are randomly distributed, the rate of energy loss by the two processes is given by the same expression:

$$\frac{dE}{dt} = -\frac{32}{9}\pi r_0^2 c \left(\frac{E}{mc^2}\right)^2 u,$$

where u in one case is the magnetic energy density, $B^2/8\pi$, and in the other case is the energy density of the background radiation.

Subject to the same limitations, the characteristic frequency of the emitted radiation depends in the same way on the electron energy: $v = v_0(E/mc^2)^2$. For the magnetic radiation, however, $v_0 = Be/2\pi mc$, while for the Compton process, v_0 is the frequency of the background radiation. It is here that a marked contrast occurs, since interstellar field strengths are on the order of microgauss, yielding v_0 on the order of a few cycles per second, while for the 2.7° background radiation, v_0 is about 10^{11} Hz, and for the optical background, v_0 is about 3×10^{14} Hz. Therefore, electrons of the same energy yield quanta about 10^{10} times more energetic by the Compton scattering than by the synchrotron process.

The *shape* of the radiated spectrum is the same in both cases. That is, electrons having a power-law spectrum, $dJ = KE^{-b}\,dE$, generate a photon spectrum (number per unit energy interval) proportional to $W^{-\alpha}\,dW$, with $\alpha = (b + 1)/2$. The validity of this assertion is limited, however, by the requirement that the scattered photons have energy small compared with that of the electrons, less than $hv_0(E/mc^2)^2$ for the highest-energy electrons, and large compared with hv_0. Thus an electron spectrum that fits a power law over a bounded interval of electron energy produces parallel spectra of photons by the two processes, over limited frequency ranges that are widely separated.

This distinct separation of the radiated spectra offers a means of analysis of both the field and the electron spectrum in space or in discrete sources, provided that one can succeed in detecting both synchrotron radiation and Compton-scattered radiation from the same source.

5. ORIGIN OF THE PRIMARY ELECTRONS

Among the primary components of cosmic rays, the electrons and positrons are uniquely sensitive to the magnetic and radiation fields in space, to the time of storage and diffusion in the Galaxy, and to the process of acceleration in the primary sources. Therefore, interest in the electron spectrum has been high throughout the history of cosmic-ray research, and has grown rapidly in the brief period in which experimental techniques have been able to distinguish the primary electron component clearly.

Early attempts to detect the electrons directly gave only upper limits of the flux (Schein, Jesse, and Wollan, 1941; Hulsizer and Rossi, 1948; Critchfield, Ney, and Oleksa, 1952). The difficulty lies not in distinguishing electrons from protons, but in detecting a very small flux of primary electrons among the very abundant secondary electrons produced in the atmosphere or in the detector itself. The presence of electrons in the Galaxy was inferred by interpretation of the galactic radio noise as synchrotron radiation (Kiepenheuer, 1950). But it was not until 1961 that primary electrons were successfully detected in high-altitude balloon flights (Earl, 1961; Meyer and Vogt, 1961).

5.1.   Background Effects

Although some studies of the electron component have been made with satellites (Cline et al., 1964; Cline and McDonald, 1968; Cline and Hones, 1968; Fan et al., 1968; Simnett and McDonald, 1969), by far the most numerous measurements have been made with balloons in the atmosphere, and it was critically important to determine a reliable correction for the electron production in the overlying air. This has been done theoretically by Verma (1967a, 1969a) and by Perola and Scarsi (1966). In addition, many experimenters have determined the atmospheric production empirically, by studying the spectrum as a function of altitude (L'Heureux, 1967; L'Heureux and Meyer, 1968; Beedle and Webber, 1968; Webber and Chotkowski, 1967; Fanselow, 1968; Rockstroh and Webber, 1969; Israel, 1969). In general the agreement between the various determinations has been good. But at energies below 300 MeV, the correction is such a large proportion of the counting rate that the possible errors are still very serious, especially in years when solar modulation seems to reduce the flux of the primary electrons.

Other sources of serious error in deducing the galactic flux, particularly at low energies, are the day–night variation of the geomagnetic cutoff, the re-entrant albedo flux, possible solar production of relativistic electrons, and uncertainties in the correction for solar modulation.

5.2.   Spectral Data

In Fig. 3, a large number of measurements of the differential flux of primary electrons have been plotted from data in the following

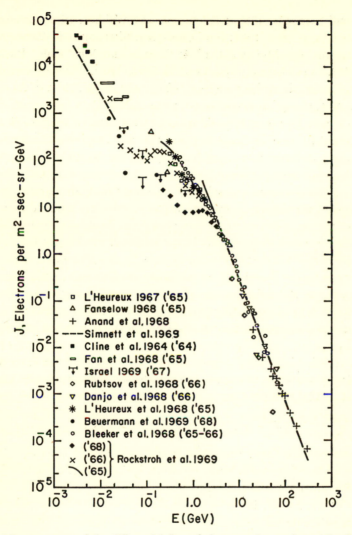

FIG. 3. Measurements of the differential flux of electrons plus positrons. In the legend, the year in parentheses is the one in which the flux was recorded: 1965 was a year of minimum solar activity. Owing to crowdedness of the figure, some points that nearly duplicate others are not shown. The data of Simnett and McDonald are represented by the dashed line instead of points. The solid line is drawn to show the fit of the high-energy points to a power-law of exponent −2.6. The solid curve suggests the probable form of the spectrum between 0.2 and 1 GeV at solar minimum.

references: L'Heureux, 1967; L'Heureux and Meyer, 1968; Rubtsov and Zatsepin, 1968; Bleeker, Burger, Deerenberg, Scheepmaker, Swanenburg, and Tanaka, 1968; Danjo, Hayakawa, Makino, and Tanaka, 1968; Anand,

Daniel, and Stephens, 1968 a; Rochstroh and Webber, 1969; Beuermann, Rice, Stone, and Vogt, 1969; Fan, Gloeckler, Simpson, and Verma, 1968; Cline, Ludwig, and McDonald, 1964; Simnett and McDonald, 1969; and Israel, 1969. In the legend printed on the figure, the number in parentheses identifies the year in which the observations were made. Most of them were made in 1965, the quietest year in the last solar cycle.

It may be noted that the measurements extend over a factor 10^5 in energy and 10^9 in differential flux and that from 10^{-1} GeV to the highest energies, the electron flux is about one percent of the primary proton flux at the same kinetic energy.

Careful examination of the data shows that throughout most of the spectrum, the results of the various experimenters agree very well as to form of the spectrum, and reasonably well even in absolute value. Indeed, for this reason some of the points in the references could not be shown in the figure, because they fall on top of each other. The interval in which agreement is best is the high-energy region, above 3 GeV. Only the points of Rubtsov and Zatsepin are significantly displaced from the others, and these seem to define a line nearly parallel to the one that fits the other points, as though the assigned energies were systematically low by a factor varying from 1.4 to 2.

In this high-energy region, from 3 to at least 300 GeV, the points also define a particularly simple and suggestive spectral shape. The line drawn on the figure represents

$$J = 116E^{-2 \cdot 6} \quad (\text{m}^2 \text{ sec sr GeV})^{-1},$$

with E expressed in GeV. The primary proton spectrum is parallel to this, displaced to the right by a factor of about 20 in energy, or upwards by a factor of about 100 in frequency.

Below 3 GeV, it is clear that the spectrum is flatter than at the higher energies. Between 0.5 and 3 GeV, the data taken at solar minimum (1965) fit a line reduced by 1.0 in slope, namely

$$J = 35E^{-1 \cdot 6} \quad (\text{m}^2 \text{ sec sr GeV})^{-1}, \quad 0.5 < E < 3 \text{ GeV}.$$

However, we have purposely displayed data taken in 1966, 1967, and 1968 as well as 1965, to show evidence that solar modulation is playing a part, below 3 GeV, in reducing the observed flux. There is experimental disagreement on the magnitude of this effect. According to L'Heureux, Meyer, Verma, and Vogt (1968), there was no detectable change (less than 60 percent) in the electron flux between 250 and 550 MeV or between 550 and

1050 MeV in the years 1960–66. But the data of Webber and Chotkowski (1967), Rockstroh and Webber (1969), Israel (1969), and Beuermann, Rice, Stone, and Vogt (1969) all show evidence of a substantial decline of the electron flux between 1965 and 1968; in some cases so severe that no difference between the total flux and that due to atmospheric secondaries was detectable in 1968.

It appears, in view of all the data, that from 30 to 300 MeV the differential spectrum in a quiet solar year is nearly flat at about 0.2 to 0.3 electrons per m²-sec-sr-MeV. When the Sun is disturbed, the flux in this energy interval may decrease by as much as an order of magnitude, and a maximum may appear at about 200 MeV, similar to the maxima that occur in the spectra of primary nuclei. If this is so, then the *galactic* flux is even higher than the flux at the Earth during quiet solar years. Not enough is known about the modulation mechanism to state the magnitude of this effect with any confidence, but according to Parker's model, with parameters matched to observed variations in the nuclear spectrum, the quiet-Sun modulation factor for low-energy particles with relativistic velocity is approximately a factor of e, so the galactic electron flux between 30 and 300 MeV is on the order of 0.5 to 1.0 electrons per m²-sec-sr-MeV.

It remains to discuss the measurements below 20 MeV. According to Cline and McDonald (1968), electrons between 3 and 12 MeV (detected with satellites outside the Earth's magnetic field) exhibit a strong 27-day semiperiodic variation, positively correlated with that of neutron monitors. The modulation of electrons is larger in amplitude than that of penetrating particles and different in its pattern: it has shown primarily short-term variations, and has not yet been studied long enough to reveal the 11-year solar cycle. Nevertheless, the phase of the magnetic storm effects suggests that the electrons are not primarily of solar origin, but are principally galactic, undergoing modulation by the Sun. It is considered likely that the average modulation factor is no bigger than in the range 30–300 MeV; in fact, there is evidence (Rockstroh, and Webber, 1969; Beuermann, Rice, Stone, and Vogt, 1969) that the modulation factor is at a maximum around 100 MeV and becomes at least as small below 10 or 20 MeV as it is above one GeV.

Thus, the spectrum exhibited between 3 and 20 MeV may be taken tentatively to be the galactic electron spectrum, possibly depressed by a small modulation factor.

The data of Simnett and McDonald (1969) fit the expression

$$J = 150E^{-1.75} \ (\text{m}^2 \ \text{sec sr MeV})^{-1}, \quad 3 < E < 20 \ \text{MeV}.$$

The data of Cline *et al.* (1964) and of Fan *et al.* (1968) agree with the same slope of the spectrum, but suggest a coefficient of 300–400 $(m^2 \sec sr \, MeV)^{-1}$ instead of 150. In view of possible modulation effects, the correct coefficient in the *galactic* spectrum could be even higher but it seems more likely that the modulation is not large and the coefficient is in the neighborhood of 200–250 per m^2-sec-sr-MeV.

5.3. Charge Ratio

As an aid in understanding the origin of the electrons, we have information not only on the shape of the spectrum, but also on the relative numbers of positrons and electrons. Such information has been given by De Shong, Hildebrand, and Meyer (1964); Hartman, Hildebrand, and Meyer (1965); Hartman (1967); Bland *et al.* (1966); Beuerman, Rice, Stone, and Vogt (1969); and Cline and Hones (1968).

Bland *et al.* observed an east-west asymmetry in the primary electron flux above 5.36 GeV, consistent with the charge ratio, $R = J^+/(J^+ + J^-)$ being zero to 0.3: i.e., with the positrons being much less numerous than the electrons. Hartman (1967) gave the following table of experimental values of R vs. electron energy (Table III).

Table III

Energy interval (GeV)	$J^+/(J^+ - J^-)$
0.038–0.087	0.49 ± 0.07
0.087–0.19	0.44 ± 0.05
0.19 –0.49	0.41 ± 0.05
0.49 –1.0	0.17 ± 0.05
1.0 –2.0	0.08 ± 0.03
2.0 –5.0	0.08 ± 0.05
5.0 –10	<0.18

Beuerman *et al.* (1969) gave a value of $R \simeq 0.3$ for the electrons and positrons between 12 and 220 MeV. At very low energy (less than 20 MeV), where the principal source of fast electrons is believed to be the knockon process, positrons would be expected to be rare, though some should originate by radioactive decay processes (Verma, 1969b). Cline and Hones (1968) sought evidence of low-energy positrons by detecting the annihilation radiation, but were only able to obtain an upper limit of the flux, which was not low enough to test these expectations.

At energies above 30 MeV, it had been thought that a principal source of the electrons and positrons would be decay of mesons, produced in space by collisions of cosmic-ray nuclei with the gas. However, as shown in Sec. 4, if that were so, positrons should be much more numerous than electrons: the ratio R should be about 0.7. Instead, we find it to be 0.5 or less, diminishing with increasing energy. At high energy, instead of the positrons being twice as numerous as electrons, the electrons are about ten times as numerous as the positrons.

Therefore a strong conclusion is possible: the electrons are directly accelerated in the sources and do not originate primarily, either in the sources or in interstellar space, as secondaries of the protons.

5.4.   Analyses of the Spectrum

Nevertheless, *some* positrons must originate in that way; and it is safe to assume (since the antiproton flux is very low, less than 10^{-3} times the number of protons) that the positrons we observe are *not* directly accelerated in the sources, but come into being *only* by way of meson-producing interactions of the nuclei. Ramaty and Lingenfelter (1968) took advantage of this principle as follows. From the measured electron spectra and the experimental ratio R of positrons to total electronic flux, they obtained the absolute spectrum of the incident positrons. This they compared with their calculation of the expected spectrum, arising from collisions of cosmic-ray nuclei with the gas in interstellar space. The excellent agreement they obtained at energies large compared with a GeV, where solar modulation is insignificant, was taken as confirmation of the basic assumptions. The calculated spectrum was then multiplied by various modulation functions, and the experimental points at the lower energies were used as a test of the validity of various forms of this function. Once the modulation function was verified, this could be applied in reverse to the total incident electron spectrum, thus obtaining the electron spectrum in galactic space. This result includes both the electrons from primary sources and the secondary electrons generated by collisions in interstellar space. The latter component, however, is known from the aforementioned calculations, which were tested against the observed positron spectrum. Finally, by subtraction of this component, they obtain the spectrum of the directly accelerated electrons.

This scheme is self-consistent and in the end yields (1) a tested solar modulation function for electrons, (2) a tested spectrum of secondary positrons and electrons originating in interstellar space, and (3) the spectrum of electrons accelerated in the primary sources. As it was applied, the scheme revealed no fallacies. However, the precision and energy range of the available data were not sufficient to permit precise deductions of these three functions. This method remains, therefore, an elegant approach to the analysis of the electron spectra, but one that must await more precise spectral data for its successful application. Especially essential are data on the positron flux at energies below 100 MeV.

Without need of such an elaborate approach, one can infer directly (from the experimental positron fraction, and from the expected value of this fraction for meson decay products) that at energies above 0.5 GeV, where $R < 0.2$, at least 70% (and probably 80–90%) of the total electron component is directly acelereated, while less than 30% (diminishing to 10% or less as the energy rises) is of secondary origin. Between 50 MeV and 300 MeV, the fraction that is of secondary origin is larger, on the order of one-half. Below 20 MeV, the flux is probably mainly of secondary origin, principally through the knockon process.

Besides the abovementioned calculations of Ramaty and Lingenfelter, there have been published a number of other analyses of the production of electrons in the Galaxy, and of what happens to them as a result of diffusion, leakage, and energy loss. Among the recent articles relevant to our discussion are the following: Perola, Scarsi, and Sironi, 1967, 1968; Ramaty and Lingenfelter, 1966b; Brunstein, 1965; Abraham, Brunstein, and Cline, 1966; Krimigis and van Allen, 1967; and Jones, 1963, 1965. Articles that particularly emphasize the expected distortion of the spectrum by the energy losses that increase as E^2 include: Daniel and Stephens, 1966; O'Connell, 1966; Ramaty and Lingenfelter, 1966a; Cowsik, Pal, Tandon, and Verma, 1966; Verma 1967b; Shen, 1967, 1969; Anand, Daniel, and Stephens, 1968a; Jokipii and Meyer, 1968; and Encrenaz and Partridge, 1969.

Here we would like to carry out some very simplified calculations related to the origin of the different parts of the electron spectrum, not seeking great accuracy, but only to reveal the basic considerations.

5.5.   The Knockon Component of Low-Energy Electrons

Let $S(x)$ be the probability that a produced particle travels a path exceeding x before diffusing out of the region from which it can reach the Earth. We will assume the mean value of x, called $\bar{x}$, is 3 to 4 g/cm². as deduced for cosmic-ray nuclei from the abundance of spallation products. $S(0) = 1$ and S is monotonically decreasing. Its exact form matters very little, and for simplicity we assume a linear decrease, $S = 1 - x/3\bar{x}$. Let $J_p(\gamma)\,d\gamma$ be the primary spectrum of the nuclei that produce the knockon electrons, γ being the Lorentz factor. $J_p = J_0\gamma^{-b}$ with $b \simeq 2.5$ and $J_0 \simeq 1.5 \times 10^4$ per m²-sec-sr. Let $dP(E')/dx$ be the probability per g/cm², that a primary produces a knockon electron of energy E' in dE'. For electrons detected at energy E, we take $E' = E + ax$ with $a \simeq 4$ MeV per g/cm², since we are dealing with low-energy electrons (1–50 MeV), that lose energy mainly by ionization. The knockon flux is then given by

$$dJ(E) = \int_x \int_\gamma S(x)\, J_p(\gamma)\, d\gamma\, \frac{dP(E')}{dx}\, dx.$$

For protons in a hydrogen target,

$$\frac{dP}{dx} = 0.15\,(2m_ec^2)\,\frac{dE}{(E + ax)^2}\left[1 - \frac{E + ax}{2\gamma^2 m_e c^2}\right].$$

After inserting the expressions for S, J_p, and dP/dx in the equation for $dJ(E)$ the integration is trivial. The result is

$$\frac{dJ(E)}{dE} = \frac{0.15 J_0}{a}\,\frac{4}{(b + 1)(b^2 - 1)}\left(\frac{2m_e c^2}{E}\right)^{(b+1)/2}$$

$$\times \left\{1 - \frac{2}{b-1}\,\frac{E}{3a\bar{x}}\left[1 - \left(\frac{E}{E + 3a\bar{x}}\right)^{(b-1)/2}\right]\right\}.$$

For $E \ll 3a\bar{x} \simeq 45$ MeV, the bracketed expression reduces to 1.0 and we have

$$\frac{dJ}{dE} = 127 E^{-1.75} \text{ per m}^2\text{-sec-sr-MeV (with } E \text{ in MeV)}.$$

We should also take into account the heavy nuclei in the cosmic rays, since the knockon cross section varies as Z^2. This raises the result by a factor of about 1.5, yielding

$$\frac{dJ}{dE} \simeq 190 E^{-1.75}\ (\text{m}^2\ \text{sec sr MeV})^{-1}.$$

This is in excellent agreement with the low-energy data shown in Fig. 3. It may be remembered that the data of Simnett and McDonald (1969) fitted $dJ/dE = 145 E^{-1.75}$ in these units, while the points of Cline, Ludwig, and McDonald and those of Fan, Gloeckler, Simpson, and Verma fell a little higher. The consistency supports the evidence, quoted above, that solar modulation effects become very small at low energies.

It may be noted that the result of the above calculation is independent of $\bar{x}$ and depends only on the rate of ionization loss and on the normalization and slope of the primary nuclear spectrum. Therefore any substantial discrepancy would be disturbing.

At energies above 10 MeV, the second term in the bracket starts to be appreciable and the spectrum gradually grows steeper. When $E \gg 3a\bar{x}$, it is steeper by one power of E: $dJ/dE \propto E^{-2.75}$. The data in Fig. 3 contain an indication of this steepening, which is due to escape of electrons from the system. Because of the steepness of the knockon spectrum and the influence of meson decay modes of electron origin, the knockon electrons are negligible at energies above 30 MeV.

5.6.   The Plateau Region

Between 30 and 300 MeV the electron spectrum is rather flat and this suggests a decay mode of origin. Even muons at rest give 35 MeV on the average to the decay electron, and the existence of a threshold velocity of the center of mass, as well as flatness of the nuclear spectrum near the threshold energy, results in large Doppler factors; hence the electron spectrum should have a peak near 100 MeV and be fairly flat over several hundred MeV before starting the steep descent which is characteristic at high energies.

The cross section is about 30 millibarns, yielding a mean free path $\lambda \cong 57$ g/cm^2. Each reaction gives about one charged pion on the average. Hence the total production of electrons and positrons is roughly

$$J = \frac{3}{2} \frac{\bar{x}}{\lambda} \int J_p(\gamma) \, d\gamma \simeq \frac{1}{10} \quad \text{(total nuclear flux)}$$

$$\simeq 400 \text{ per m}^2 \text{ sec sr}.$$

Since most of these electrons are distributed between zero and 400 MeV in energy, we estimate a differential plateau flux, dJ/dE, of about one per m^2-sec-sr-MeV, most of these particles being positrons. Solar modulation probably reduces this flux by a factor of two or three at the Earth.

This calculation is very crude, simply because a more refined one is quite difficult and doesn't make much difference. Ramaty and Lingenfelter (1966b) also predict plateau values of 1 to 2 (m^2 sec sr MeV)$^{-1}$ without modulation, and 0.7 in the same units after taking solar modulation into account. Krimigis and van Allen (1967) get a plateau value of 1.7 unmodulated). The reader can refer to these papers and the ones by Perola,

Scarsi, and Sironi for careful treatments of the problem. But it should be apparent from our estimate that in this part of the spectrum the result is proportional to the mean amount of matter traversed by the primaries, $\bar{x}$, and is not very sensitive to anything else.

The data in this part of the spectrum, during a quiet solar year, indicated a plateau level of approximately 0.3 electrons per m²-sec-sr-MeV, with no more than half of the flux being due to interstellar secondaries. So we have a slight discrepancy between predicted flux values of about 0.4 to 0.8 in these units, and an observed flux of no more than 0.2 that can be attributed to interstellar meson production. However, the problem of solar modulation is so complex and so little understood, that a discrepancy of this small magnitude is not very startling.

5.7.   The Spectrum above 2 GeV

Full absorption of the primary cosmic rays in the Earth's atmosphere has been found to yield production spectra as follows:

$$\frac{dJ}{dE} = KE^{-8/3} \text{ (m}^2 \text{ sec sr BeV)}^{-1}$$

with K = 1560 for charged pions
    = 590 for photons
    = 550 for neutrinos
    = 200 for electrons.

This result has been obtained in many experimental studies of decay products of the pions in the atmosphere: particularly the muons and photons. See, for example, Pine, Davisson, and Greisen, 1959. The origins of the various secondaries are so closely coupled through the decay modes and charge symmetry, that a measurement of any one type of secondary determines the production spectrum of all the others. The value of the exponent is also understood in terms of the exponent, b, of the primary spectrum and the variation of the multiplicity of pion production with the 1/4 power of the primary energy. It follows from this that the exponent of the pion spectrum is $-(4/3)(b - 1/2)$ which is $-8/3$ when $b = 2.5$. The decay products all have the same spectral exponent as the pions in the high-energy domain.

The above expression is valid for pions between 5 and at least 400 GeV, and hence for decay electrons between 1 and at least 100 GeV.

The absorption length of the primaries is double the interaction length, and the latter (in hydrogen) is about 57 g/cm². Hence, with

$\bar{x} \simeq 4 \text{ g/cm}^2$, the spectra generated in interstellar space should be like those produced in the atmosphere except for a reduction of the coefficients by a factor of about 0.05. Thus we arrive at a predicted spectrum of secondary electrons and positrons:

$$\left(\frac{dJ}{dE}\right)_{\text{sec}} \simeq 10 \, E^{-8/3} \quad (\text{m}^2\text{-sec-sr-GeV})^{-1}.$$

The measured flux at high energies was consistent with $(dJ/dE)_{\text{total}} = 117E^{-2.6}$.

Thus, these calculations suggest that only about nine percent of the high-energy electron flux should be of secondary origin through pion production in interstellar space. Exactly the same conclusion was indicated by the data on the relative proportion of electrons and positrons.

5.8.   Distortion due to Energy Losses

Since the amount of matter traversed by primary electrons before reaching the Earth is only a few g/cm², ionization losses are insignificant except for electrons below 50 MeV, and interstellar energy loss by bremsstrahlung is never a major factor (the radiation length is 63 g/cm² in hydrogen and 93 g/cm² in helium). However, the losses by synchrotron radiation and inverse Compton effect are proportional to E^2:

$$\frac{dE}{dt} = - \, 10^{-16} \, uE^2 \text{ GeV per second,}$$

with E in GeV and u the total energy density, in eV/cm³, of the magnetic field and the background radiation. For electrons of any particular mean age, τ, there is always a characteristic energy E_c above which these losses become quite drastic. E_c is defined by

$$\frac{dE}{dt}(E_c) = - \, \frac{E_c}{\tau};$$

hence $E_c = 10^{16}/u\tau$ with u in eV/cm³, τ in seconds, and E_c in GeV.

In fact, integration of the energy loss for a particle that has energy E_0 at $t = 0$ yields

$$\frac{1}{E(t)} = \frac{1}{E_0} + \frac{t}{\tau E_c};$$

hence E_c is the maximum energy of any particle that has survived for a time τ, even if the injection energy is unlimited. If particles are injected with a spectrum $dJ_0 = KE^{-b}\, dE$, then after a time t, the spectrum will have been reduced to

$$dJ(E) = KE^{-b}\, dE \left(1 - \frac{t}{\tau} \frac{E}{E_c} \right)^{b-2}.$$

If particles are *continually* injected at the rate dJ_0 per unit time, the equilibrium spectrum that is attained depends on the distribution of age among the particles reaching Earth. This depends in turn on the location of the sources and the solution of the diffusion and leakage problem. Examples of such calculations for specific models are in the articles by Shen (1967) and Jokipii and Meyer (1968).

In general, however, one can see that the effect on the spectrum is very little at energies less than E_c, while for energies large compared with E_c we receive only those particles that reach us in a time $t < \tau E_c/E$. If the age distribution is flat for small values of t, this means that the spectrum has been cut by a factor proportional to E^{-1}; hence its exponent has been increased in magnitude from b to $b + 1$ at energies large compared with E_c.

Let us now look at the expected magnitude of E_c. The particles reaching Earth have penetrated, on the average, 3 to 4 g/cm² of interstellar gas. If they do this within the disk of the Galaxy where the gas density is about 1.6×10^{-24} g/cm³, the mean time, τ is about 7×10^{13} sec, hence $E_c \cong 140/u$ GeV. Now the magnetic energy density in this region (taking $B = 6 \times 10^{-6}$ gauss) is 0.9 eV/cm³; the optical energy density is 0.3 eV/cm³; and the 2.7° background is 0.25 eV/cm³. However, there is great uncertainty about the amount of infrared. Detected infrared galaxies add comparatively little to u, but Shivanandan, Houck, and Harwit (1968) have discovered an isotropic component outside the atmosphere, having an energy density of 13 eV/cm³ in the wavelength range 0.4–1.3 mm. [A second rocket flight (Houck and Harwit, 1969) has confirmed the existence, the intensity, and the isotropy of this radiation.] It is not yet known whether this radiation is a local phenomenon (case I) or universal (case II), so we must treat these possibilities separately. In case I, u is about 1.5 eV/cm³ and E_c (for disk storage) is about 100 GeV. In case II, a lower limit of u is 15 eV/cm³ and E_c is less than 10 GeV.

In the halo of the Galaxy we take $B \cong 2.5 \times 10^{-6}$ gauss, the optical density as 0.1 eV/cm³, and τ as 6×10^{15} sec. In case I (negligible infrared), $u \cong 0.5$ eV/cm³ and $E_c = 3$ GeV, while in case II, $u \geqq 13.5$ eV/cm³ and $E_c = 0.1$ GeV.

For electrons in intergalactic space we take $\tau \cong 2 \times 10^{17}$ sec (based on the universal expansion), the magnetic energy density as negligible, the optical density as 10^{-2} eV/cm^3, and the 2.7° radiation as 0.25 eV/cm^3 as before. In case I (including only infrared from observed galaxies)

$$u \cong 0.3 \text{ eV/cm}^3 \quad \text{and} \quad E_c \cong 0.17 \text{ GeV}$$

while in case II, $E_c \cong 0.004$ GeV = 4 MeV!

Now what can we conclude from the fact that the electron spectrum is parallel to the proton spectrum with a slope of -2.6 up to energies beyond 150 GeV? There are two possibilities. One is that the detected electrons originate near us in the Galaxy and traverse even less than 4 g/cm^2 on the average before reaching the Earth. Then the steepening of the spectrum by energy losses would occur near or beyond the upper energy limit of present observations. In case I (little infrared), the mean path length would not have to be *much* less than 4 g/cm^2—hardly a factor of 2; but in case II (universal infrared) the mean path length would have to be less than 0.3 g/cm^2.

[In this connection, it is interesting that an analogous problem has arisen concerning the low-energy part of the *nuclear* spectrum, which should also be distorted by energy losses. The similarity of the low-energy spectra of nuclei with different values of Z has indicated that a substantial part of the arriving flux has traversed only a small fraction of one g/cm^2 of interstellar gas.]

The second possibility concerning the electron spectrum is that E_c lies as low as 3 GeV or possibly even lower, where changes in slope of the spectrum are in fact observed. These changes of slope *could* have other causes, but one of the causes could be the energy losses. In this case, the injection spectrum of electrons at the sources would have to be flatter by one power of E than the spectrum observed at Earth at energies above 3 GeV. The production spectrum of electrons would therefore have to be flatter than that of the nuclei, and the observed parallelism would have to be an accident. Moreover, at very high energies the production of electrons would have to exceed that of nuclei, and would place a heavy burden on the energy requirements in the sources. All of these implications seem to me to be highly implausible.

Moreover, we would have to conclude that the spectrum of electrons existing in the halo of the Galaxy is strongly depleted in comparison with the spectrum near the Earth. Then it would be difficult to account for the diffuse nonthermal radio background, which is believed to be caused by halo electrons.

5.9.   Galactic Radio Emission

Radio receivers can resolve numerous external galaxies as well as discrete sources in our own Galaxy, and from the signal strengths a luminosity function is derivable, which can even be extrapolated to include signals from the galaxies that are too weak or distant to detect individually. The summed signals from all the discrete sources, however, are insufficient to account for the total energy received. The residue includes a strong, isotropic microwave signal with a Planck frequency distribution corresponding to $T = 2.7\,°K$—the universal thermal background radiation. This dominates the spectrum at frequencies above 10^9 Hz. At lower frequencies the spectrum is nonthermal in character, and the intensity varies with galactic latitude and longitude, being strongest in the plane of the Galaxy, and within this plane being strongest towards the galactic center.

This nonthermal radiation is attributed to magnetic bremsstrahlung of relativistic electrons in a magnetic field that exists both in the disk of the Galaxy and in a roughly spherical halo surrounding the disk. The existence of such a halo has been questioned, but the radio flux at high galactic latitudes is too intense to be due to electrons in the disk, and is more easily accounted for in terms of a halo than by isotropic radiation from intergalactic space (see, e.g., Felten, 1966; Anand, Daniel, and Stephens, 1968b).

Below a frequency of 10^7 Hz, the radio spectrum is modified by interstellar absorption by ionized hydrogen, and above 10^9 Hz the nonthermal radiation is obscured by the thermal background. So the frequency range in which the spectrum can be studied is limited to 10^7–10^9 Hz. The fields responsible for the emission apparently vary from about 2.5 microgauss in the halo to 5–20 microgauss in different parts of the disk. Therefore the electrons principally contributing to the radiation are in the energy range 300 MeV–10 GeV.

Numerous analyses have been made of the spectral and directional distribution of the radiation, in terms of the electron spectrum and field strength in different parts of the disk and halo. See, for instance, Felten, 1966; Anand, Daniel, and Stephens 1968c, and Webber, 1968. The radio power spectrum does not follow a simple power law, but has a negative spectral index that increases from about 0.3 at 10^7 Hz to 0.8 at 10^9 Hz. Because of this, a self-consistent scheme of analysis is possible in which both the field strength and the electron spectrum can be deduced. The conclusions are (1) that the average field strength is as mentioned above: 2–3 μG in the halo and 5–10 μG in the disk (except towards the center where the field is stronger); and (2) that the electron spectrum is similar throughout the

5*

Galaxy (including the halo) to the spectrum at the Earth, except for some reduction of the latter at the lowest energies by solar modulation.

The crucial point related to our previous discussion of the electron spectrum is that the spectrum in the halo seems to be very little steeper than that in the galactic disk.

It is quite difficult to reconcile this result with the universal existence of the strong infrared radiation observed near the Earth by Shivanandan, Houck, and Harwit (1968). Even without this source of energy loss, the electron spectrum in the halo should be steeper than the local spectrum unless the mean storage time in the halo is reduced from 2×10^8 years to 1×10^8 years or less; but if the infrared were present, the electrons above one GeV in the halo should be so depleted in number that their synchrotron radiation would be undetectable; and its spectrum should also be considerably steeper than that of the disk radiation.

5.10.   Summary

In conclusion, the data on the electron spectrum argue strongly against the universal existence of infrared radiation with energy density greatly exceeding that of the 2.7° background. The complete lack of evidence for steepening of the electron spectrum by energy losses due to synchrotron radiation and Compton scattering indicates that the mean age of the electrons is somewhat less than has been believed in the past. Namely, it is no more than about one million years in the disk of the Galaxy and 10^8 years in the halo. But if the strong infrared radiation were universal, the mean age of the electrons would have to be lower than these values by two orders of magnitude, and the rate of electron production in the sources would have to be correspondingly higher. There is no reason to expect the storage time of nuclei in the Galaxy to be any longer than that of electrons of the same rigidity. Therefore the required rate of energy conversion to cosmic rays would be raised to a level very difficult to account for, the abundance of nuclear spallation products would no longer be understood, and intergalactic space would be almost as densely populated with cosmic rays as is the Galaxy. Altogether, it seems more likely that the infrared detected at rocket altitudes is due to some obscure effect occuring in the outer atmosphere of the Earth or in interplanetary space.

The electron flux is so low at energies above 100 GeV that its detection is difficult. But we look forward, anyway, to detailed experimental studies in the energy range 10^{11}–10^{15} eV, in order to resolve the questions that remain about the electron acceleration process and the time spent in transit between the sources and the Earth. In this high-energy range, the observations will be extremely sensitive to the energy losses in space and also in the sources themselves. If the predicted steepening of the spectrum is not observed, this will imply an enormous outpouring of energy in cosmic-ray sources, and that cosmic rays at their origin are predominantly electrons rather than nuclei.

6.   COSMIC GAMMA RAYS

The search for cosmic gamma radiation was stimulated a decade ago by an article by P. Morrison (1958). Since then, the processes of origin have been outlined in a number of reviews: Pollack and Fazio (1963); Ginzburg and Syrovatskii (1964); Hayakawa and Matsuoka (1964); Gould and Burbidge (1965, 1967); Garmire and Kraushaar (1965); Fazio (1967); and Lüst and Pinkau (1968). The state of experimental progress has been summarized fairly recently by Greisen (1966) and Duthie (1968). All of these articles, however, preceded the magnificently successful observations of Clark, Garmire, and Kraushaar (1968) as well as the discovery of pulsars; so one must not expect the story they present to be complete.

6.1.   Detection Methods and Difficulties

In the first decade of experimental efforts to detect cosmic gamma rays the results were largely negative, in contrast to the quick success and rapid accumulation of data in the experimental field of x-ray astronomy. The difference is not due to a disadvantage in *energy* flux, but due to the contrast in number of photons, and to associated problems regarding background, shielding, and collimation.

In fact, the energy flux appears to be distributed with respect to quantum energy approximately as dE/E all the way from the soft x-ray range (less than one kilovolt) to very hard gamma rays (at least 200 MeV), implying equal amounts of energy in equal logarithmic intervals, or successive decades, of the energy. But the *number* of photons above one keV in the x-ray range is about 15 per cm^2-sec-sr, which is 100 times the charged cosmic-ray flux at medium latitudes; while the number of gamma rays above 100 MeV, averaged over direction in space, is only about 2×10^{-4} per cm^2-sec-sr, a thousand times *less* than the primary charged particle flux. Since the charged particles generate secondary gamma rays locally with high efficiency, it is obvious that the problem of avoiding spurious background is much more severe in trying to detect gamma rays than in measuring the x-rays.

There is also the simple but fundamental problem of size of apparatus and time required to make an observation. For instance, the gamma-ray detector on OSO-III (Clark *et al.*, 1968) has an effective aperture (area × efficiency × solid angle) of about 0.5 cm²-sr. Therefore it detects gamma rays at the low rate of one per three hours while pointing away from the Earth: only a few per day on the average because of the time lost when the Earth is in the field of view. Even with much larger detectors, rockets are not a useful vehicle for getting above the atmosphere because a few minutes spent above the atmosphere is quite inadequate to provide any significant data.

X-ray detection also has the advantage that thin honeycomb cells or fine wires can be used to provide collimation of the photons. Gamma rays, on the contrary, would require screens many inches thick, and inter-actions of charged cosmic rays in the collimating materials would generate a hopelessly copious background, overwhelming the small flux of primary gammas that enter through the defined opening.

Probably the most difficult energy range in which to carry out gamma-ray astronomy is from 1 to 10 MeV. Here the usual methods make use of shielded crystal scintillators and/or solid state detectors. But the secondary electrons have very short range and undergo severe scattering, while the primary gammas have very long absorption lengths. Hence the detectors have been essentially omnidirectional; no detector incorporating both high efficiency and good directionality in this energy range has yet been designed. At slightly lower energies, 0.1 to 1 MeV, however, the shorter gamma-ray absorption length makes possible a highly effective directional instrument, in which the detecting element is surrounded by an active collimator (a second scintillator), that vetoes charged cosmic rays entering in all directions, as well as gamma rays in directions outside a certain aper-ture, defined by drilled holes or a single well-type hole in the active colli-mator. Such instruments have been developed and applied by L. Peterson at La Jolla and R. C. Haymes of Rice Institute (see, e.g., Haymes and Craddock, 1966). With increasing energy above one MeV, the directional response of such instruments rapidly degenerates.

Above 50 MeV, the difficulty in defining an aperture with colli-mators persists, and yet the most significant data have been recorded with instruments of this type: scintillators surrounded by collimating shields [Kraushaar, Clark, Garmire, Helmken, Higbie, and Agogino (1965); Clark,

Garmire, and Kraushaar (1968)]. However, the angular definition provided by those instruments was only 15°, which severely limits the conclusions that can be drawn about the identity of the sources of the radiation.

The most popular detector of gamma rays above 50 MeV has been the spark chamber [Frye and Smith (1966); Frye and Wang (1968, 1969); Ögelman, Delvaille, and Greisen (1966); Fazio, Helmken, Cavrak, and Hearn (1968); Fazio and Helmken (1968); Fichtel (1968); Fichtel, Cline, Ehrmann, Kniffen, and Ross (1968); Fichtel, Kniffen, and Ögelman (1969); May and Waddington (1969); Duthie (1968)]. With this instrument one takes advantage of the fact that the photon converts into an electron–positron pair that have substantial range, and initially maintain the gamma-ray direction within an angle of mc^2/E radians, m being the electron mass and E the gamma-ray energy. Until 1970 at least, however, no spark chamber will have been flown in a satellite; and at balloon altitudes the atmospheric background has obscured the *diffuse* primary flux, and prevented detection of discrete sources that produce a flux at the Earth less than 10^{-5} photons per cm²-sec.

With increasing energy beyond 10 GeV, the anticipated primary photon flux is so low that spark chambers cannot be made large enough to have any promise of success. At still higher energies, however, another method becomes possible, namely detection of the showers which the photons generate in the air. With this method one has only indirect information that the primary is a gamma ray. The most convincing evidence is provided by angular resolution: primary charged particles are smoothly distributed in direction, and only gamma rays (since neutrinos interact with negligibly low efficiency) can yield peaks in the angular distribution at the coordinates of discrete sources or distinct features of the Galaxy. The advantage of shower detection is that the effective area of the detector is the area of the shower, on the order of 10^9 cm². Therefore very many events can be recorded, and even peaks corresponding to less than one percent of the smooth background in the resolved cone of angles can be discerned.

The favored means of detecting air showers with good angular resolution has been to utilize large focussing mirrors at ground level, and record the pulses of Čerenkov light produced by the showers in the atmosphere [Chudakov, Dadykin, Zatsepin, and Nesterova (1962); Fruin, Jelley, Long, Porter, and Weekes (1964); Charman *et al.* (1968); O'Mongain *et al.* (1968); Fegan *et al.* (1968); Fazio, Helmken, Rieke, and Weekes (1968a, b, c)]. The small angle of emission of Čerenkov light makes possible an angular resolution of about one degree (although most of the experiments did not actually achieve resolution better than two degrees). The threshold gamma-ray energy needed to provide an optical pulse exceeding the noise due to starlight has usually been about 5×10^{12} eV. However, by use of a larger telescope and faster electronics, Fazio *et al.* (1968c) have been able to reduce the threshold to 4×10^{11} eV.

A second way of distinguishing air showers due to primary gamma rays is by the virtual absence of muons in them. This technique has been applied by the cosmic-ray group at Lodz [Gawin *et al.* (1965); Catz *et al.* (1967)] and in the international air shower experiment in Bolivia [Suga *et al.* (1963); Hasegawa *et al.* (1965)]. An unusual paucity of muons does not guarantee that a shower was initiated by a gamma ray, but it makes the probability much higher. Only about 10^{-3} of the air showers of a given "size" (number of particles) show this lack of muons, so the selection technique at least rejects a very large proportion of the showers that are due to charged primaries. An effective threshold energy for application of this selection technique is about 10^{14} eV, since the shower must be large enough to contain many muons if it is normal in character. Unfortunately, this is also the energy at which the Universe becomes opaque to high-energy gamma rays, as explained in Sec. 4. Nevertheless, some gamma rays above 10^{14} eV may arrive from sources in our Galaxy or its nearest neighbors, and these gammas may be distinguishable by applying the joint criteria of absence of muons *and* exhibition of directional features.

6.2. Observations of Isotropic Gamma-Ray and X-Ray Background

We now turn to a summary of information that has been accumulated by the techniques outlined above. In the quantum energy range ordinarily referred to as gamma rays (energy high compared with 10^5 eV) there are not many experiments that have produced more than upper limits of the flux. To be precise, there are just the measurements between 60 keV and one MeV made on two flights of Ranger spacecraft by Arnold, Metzger, Anderson, and van Dilla [Arnold *et al.* (1962); Metzger *et al.* (1964)], and the measurements in the neighborhood of 100 MeV made on the Explorer XI and OSO-III satellites by Kraushaar, Clark, Garmire, Helmken, Higbie, and Agogino [Kraushaar *et al.* (1965); Clark *et al.* (1968)].

Neither of these experiments had good angular resolution. However, the Ranger apparatus had enough variation of sensitivity with direction to have been able to see a gross change of flux, and saw none. The MIT apparatus depended critically on the angular resolution that it had ($\pm 15°$), because the vehicle remained near the Earth, and the albedo flux from the atmosphere was 20 to 100 times the flux from space. When this instrument was pointed away from the Earth, it observed an enhanced flux in the plane of the Galaxy, but also a flux at high galactic latitudes which seemed to be distributed uniformly over the celestial sphere.

Thus, the gamma rays, like the primary x-rays, seem to have two components, one associated with features of, or objects in, the local Galaxy, and the other distributed isotropically—hence probably of extragalactic origin and cosmological significance. In fact, at x-ray energies below a kilovolt, the extragalactic origin has been substantiated by a departure from isotropy with a *minimum* at low galactic latitudes, indicating absorption by the interstellar gas in the Galaxy [Henry *et al.* (1968); Baxter, Wilson, and Green (1969)]. This does not prove that the flux at higher energy, where the absorption is negligible, is also of extragalactic origin. But in view of the extreme anisotropy of the Galaxy and of the potential local sources within it, a roughly uniform flux of penetrating radiation, like that which has been observed, is almost certainly of external origin.

In absolute intensity and spectral shape, there is no observable discontinuity in the background spectrum between 1.5 keV and one MeV, a range of energy replete with observations. At most, there are only small changes in slope of a smooth curve. Moreover, an extrapolation of this curve comes close to fitting the observation at 100 MeV. Therefore it looks as if the background from 10^3 to at least 10^8 eV represents a single phenomenon. Hence, we summarize the x-ray and gamma-ray data together at this point, instead of reserving the x-ray background for presentation in Sec. 7.

The data available prior to 1967 have already been surveyed in a review article by R. J. Gould (1967). Briefly, the data between 0.05 and 1.0 MeV, including the Ranger data of Metzger *et al.* (1964), are an excellent fit to the expression

$$dJ/dE = 100E^{-2.3} \text{ photons/cm}^2\text{-sec-sr-keV (with } E \text{ in keV)}$$

$$= 0.013E^{-2.3} \text{ photons/cm}^2\text{-sec-sr-MeV (with } E \text{ in MeV)}.$$

The data presented by Gould (1967) as well as more recent data [Bleeker *et al.* (1968)] agree in indicating that at the lower end of the above range, the spectrum is slightly steeper. Gould estimates the exponent to be 2.5, while Bleeker *et al.* (1968) find that their data in the range 20–180 keV fit the expression

$$dJ/dE = 135E^{-2.4} \text{ photons/cm}^2\text{-sec-sr-keV (} E \text{ in keV)}.$$

However, at still lower energies the spectrum certainly grows flatter. In units of photons/cm²-sec-sr-keV, Seward *et al.* (1967) give

$$dJ/dE = 9E^{-1.6} \text{ from 4 to 40 keV};$$

Gorenstein, Kellogg, and Gursky (1969) give

$$dJ/dE = 12.4E^{-1.7\pm0.2} \text{ from 1 to 13 keV};$$

Henry *et al.* (1968) show data that agree with

$$dJ/dE = 9.5E^{-1.4\pm0.1} \text{ from 1.5 to 8 keV};$$

and Baxter, Wilson, and Green (1969) give

$$dJ/dE = 12E^{-1.49\pm0.05} \text{ from 1.6 to 12 keV}.$$

These spectra are all in excellent agreement and merge smoothly with each other, showing no discrepancies in absolute level. So the general picture is of a smooth power-law spectrum over a wide range of intensity and quantum energy: the logarithmic slope being 1.4 around 2 kilovolts, increasing gradually to about 2.5 at 30–50 kilovolts, then decreasing to about 2.2 between 200 kilovolts and one MeV.

The isotropic flux observed by Clark *et al.* (1968), integrated over energies above 100 MeV, is $(1.1 \pm 0.1) \times 10^{-4}$ per cm²-sec-sr. If one seeks to join this value to the spectrum at one MeV by a straight-line extrapolation of the latter spectrum, the required index of the differential spectrum is 2.0 (an extrapolation from 1 MeV with a slope of 2.2 falls below the MIT point by a factor of 2.8).

Below one kilovolt, there is agreement that the spectrum steepens, but wide disagreement as to the magnitude of this effect [Bowyer, Field, and Mark (1968); Henry *et al.* (1968); Baxter, Wilson, and Green (1969)]. At the lowest energy where data are given, 0.27 keV, the reported flux varies from 270 to 2900 photons per cm²-sec-sr-keV. Part of the discrepancy lies in the correction for interstellar absorption.

It is impressive that over a factor of 10^{12} in differential flux and 10^{6} in quantum energy, a single power law agrees with all the data remarkably well, even falling in the midst of the reported values at 270 eV. The expression for this power law is

$$dJ/dE = 33E^{-2.1} \text{ photons/cm}^2\text{-sec-sr-keV} \ (E \text{ in keV}).$$

The corresponding integral spectrum, $J(> E) = 30E^{-1.1}$ (cm²-sec-sr)$^{-1}$, yields a flux of 1.5×10^{-2} (cm²-sec-sr)$^{-1}$ above one MeV, and 1.0×10^{-4} (cm²-sec-sr)$^{-1}$ above 100 MeV.

Experimental upper limits have been obtained by several methods for the integral gamma-ray flux above various energies higher than 100 MeV. Most of these limits, if not all, are probably orders of magnitude above the actual flux values, but we list the limits anyway in Table IV.

TABLE IV

Lower energy bound	Integral flux	Reference
2 GeV	2.6×10^{-4} (cm² sec sr)⁻¹	
5 GeV	1.4×10^{-4}	Anand, Daniel, and
10 GeV	0.6×10^{-4}	Stephens (1968)
50 GeV	0.3×10^{-4}	
250 GeV	2×10^{-5}	Kidd (1962)
500 GeV	3×10^{-7}	Duthie et al. (1962)
2.5 TeV	5×10^{-7}	Abraham et al. (1963)
1000 TeV	10^{-13}	BASJE [Suga et al. (1963); Hasegawa et al. (1965)]

An effort has been made to detect characteristic gamma-ray lines in the diffuse spectrum, without success. Metzger et al. (1964) gave upper limits of the intensity of the positron annihilation line (0.51 MeV) and the neutron capture line (2.23 MeV) as follows:

$$J(0.51 \text{ MeV}) < 0.0011 \text{ (cm}^2 \text{ sec sr)}^{-1},$$

$$J(2.23 \text{ MeV}) < 0.0004 \text{ (cm}^2 \text{ sec sr)}^{-1}.$$

In addition to the isotropic flux above 100 MeV, Clark, Garmire, and Kraushaar (1968) detected an enhanced flux from the galactic disk. This intensity varied with galactic longitude, being greatest in the direction of the galactic center. Because of the crude angular resolution, it is not clear whether the galactic gamma-ray flux originates, like the x-rays, primarily in discrete objects such as x-ray stars, supernovae and/or pulsars, or whether it originates in the interstellar gas. The angular width of the peak in respect to galactic latitude was apparently smaller than the width (30°) of the response function of the OSO-III detector; hence the authors reported the flux in terms of the strength of an equivalent line source in the galactic disk. The average strength of this line source was 1.6×10^{-4} photons (above 100 MeV) per cm²-sec per *radian*, summing to 10^{-3} photons/cm²-sec over all longitudes. In the approximate direction of the center of the Galaxy, the equivalent line source had a broad maximum of 5×10^{-4} per cm²-

sec-rad, while in other directions its strength seemed to vary from approximately zero to 3×10^{-4} (cm²-sec-rad)⁻¹. Ögelman (1969) has pointed out the suggestive similarity between the distribution of amplitude of the gamma-ray source and the distribution of discrete x-ray sources in the galactic disk, noting that if the x-ray flux is extrapolated with a differential index of -2.0 in the energy spectrum, it accounts excellently for the entire galactic gamma-ray flux. Nevertheless, efforts have also been made (as discussed later in this section) to account for the gamma-ray flux as a consequence of interactions in interstellar space.

In 1969, with balloon-borne detectors, two further experiments have detected the enhancement of primary gamma rays above 50 or 100 MeV in the galactic disk [Sood (1969); Valdez and Waddington (1969)]; and in one case (Sood) the variation of the flux with galactic longitude was also confirmed. These represent the first successes in detection of primary gamma rays above 0.5 MeV with balloon-borne apparatus.

6.3.   Experimental Data on Discrete Sources

Almost all of the data on discrete sources are negative, and it is easy to understand why. Directional gamma-ray detectors are only operational above 50 MeV. Let a detector have area A, angular resolution Ω (steradians), efficiency ε, and time of exposure to a particular source T. Then the number of source counts recorded is $\frac{1}{2}JA\varepsilon T$, which must be large compared with 1. Typically A has been 400 cm² or less, $\varepsilon \cong \frac{1}{4}$, and $T \lesssim 10^4$ sec, determined by the time during which a source is in view of the apparatus. The factor 1/2 accounts for the triangular variation of the response function as the source sweeps through the field of view. One sees that a source must have $J \gtrsim 10^{-5}$ (cm² sec)⁻¹ to meet this requirement.

In addition, the background flux provides a second limitation, which is usually the dominant one. Let the background flux per steradian be B; then the number of background counts per resolution element, Ω, is $BA\varepsilon\Omega T$, and its standard deviation is the square root of this product. The number of source counts must exceed this "noise" by a factor of at least 4 for detectability. This requirement imposes the condition

$$J \gtrsim 8 \sqrt{\frac{B\Omega}{A\varepsilon T}}.$$

Typically, $B \cong 4 \times 10^{-3}$ (cm² sec sr)⁻¹ at balloon elevations, $\Omega \cong 10^{-2}$ (i.e., 3° resolution), and with A,ε,T estimated as above, we find that discrete sources cannot be detected unless $J > 5 \times 10^{-5}$ (cm² sec sr)⁻¹ at energies above 50 MeV. Some instruments have been designed to reduce this limiting flux a little, but none by a large factor. And nature has not been kind enough to provide such strong sources.

The above limitations are not permanent. The background flux (above 100 MeV) in space is lower than the value at balloon elevations by a factor of about 40; the solid angle Ω can be reduced by an order of magnitude or more; the area A can be increased by an order of magnitude; and with an oriented satellite, the time of observation, T, can be increased by two orders of magnitude. Therefore the minimum detectable flux (at energies above 50 MeV) can be reduced below 10^{-7} per cm²-sec. But the slowness of our scientific space program and recent cutbacks in funding have postponed the date when this can be attained to at least 1972, and possibly much longer.

Meanwhile, it is possible that one or more sources is close to the previous limits of detectability, and will become discernable with less ambitious improvements in balloon-borne apparatus: developments in progress can reduce the limit of detectability to 10^{-6} (cm²-sec)$^{-1}$. This is also the approximate threshold of detectability for the Goddard wire-grid spark chamber [Fichtel *et al.* (1968)] which is scheduled to be launched in 1971 in a Small Astronomy Satellite. There is also promise in further searches with the Čerenkov air shower technique. Current instrumental development [Fazio, Helmken, Rieke, and Weekes (1968 *a*, *b*, *c*)], leading to a reduction in the detectable gamma-ray energy towards 2×10^{11} eV, may allow the flux from some sources to exceed the detectable limits. The possibility that sources of high-energy gamma rays are pulsing, as was discovered for at least one x-ray source, and temporarily appeared to be the case for high-energy gammas [Charman *et al.* (1968); O'Mongain *et al.* (1968)], should be further investigated. If true, it will not only be exciting in its implications, but will also be an aid in the detection of weak sources.

Meanwhile, one can list very briefly the discrete sources that have been identified. There is one near the galactic center, observed by Clark, Garmire, and Kraushaar (1968) and confirmed by Sood (1969). This may not be a single source, but a complex of discrete sources or even intense emission from interstellar space. The total flux within 15° of the galactic center, above 100 MeV, appears to be 2.4×10^{-4} (cm² sec)$^{-1}$ according to the OSO-III measurements, implying either a diffuse flux in excess of 10^{-3} (cm² sec sr)$^{-1}$ in much of this region, or individual sources, each producing on the order of 10^{-4} photons per cm²-sec at the Earth.

Secondly, along the galactic equator, OSO-III detected an average equivalent line source of 1.6×10^{-4} photons per cm²-sec-radian, varying with galactic longitude from zero to double this value. The belt included in the analysis was 30° wide. Therefore the *average* flux in this belt was 3×10^{-4} photons per cm²-sec-sr. In view of the distribution of matter within the disk, however, the flux must be non-uniform within the 30° belt, being at least 1×10^{-3} per cm²-sec-sr near its center and approaching 1×10^{-4} near its edge. If Ögelman's suggestion (1969) is right about the gamma rays being an extension of the x-ray spectra of point sources, the galactic belt must be populated with some tens of discrete sources, some of which produce flux levels at the Earth of 5×10^{-5} photons per cm²-sec.

It is within the capability of existing instruments to check these observations, but a close examination of the publications reveals that the negative results of many of the spark-chamber balloon flights are too crude to have constituted a significant test. Unfortunately, there have been very few successful flights in the southern hemisphere, that would get a good view of the galactic center.

The observations of Fichtel *et al.* (1968, 1969), however, seem to contradict the OSO-III results to some degree. Regarding the galactic center as a potential point source, they were able to set an upper limit of 4×10^{-5} (cm² sec)$^{-1}$ to the number of photons above 30 MeV. In the 1969 publication, they analyzed the same data in terms of a potential line source. The errors were then so large that the disagreement with the OSO-III data was not convincing, but the results favored an intensity at least two times smaller than that given by Clark *et al.*

A slightly stronger contradiction appears in the spark-chamber observations of Frye and Wang (1969). They surveyed most of the northern hemisphere and found no sources yielding as many as 2×10^{-5} photons per cm²-sec above 50 MeV, or as many as 1×10^{-5} above 150 MeV. In particular, they examined two strips of the sky along the galactic disk, one in the Cygnus region and one near the galactic anticenter. The strips were 6° wide, defined by $|b_{II}| < 3°$, in regions where Clark *et al.* had seen an equivalent line source of 1.5×10^{-4} photons per cm²-sec-radian above 100 MeV, using a strip five times as wide. Frye and Wang found an average flux less than 20% above background in these strips, implying a line source of less than 3×10^{-5} photons per cm²-sec-radian. Since the strips are of unequal width, the comparison is not direct, but it seems as if either the analysis of Frye and Wang yields upper limits too small by a factor of two or three, or else the calibration of the OSO apparatus yields intensities that are too high by a similar factor. It is premature to be disturbed by such a small discrepancy. Clearly what is needed is more data, even with instruments having present capabilities, especially involving flights in the southern hemisphere.

Particular interest has centered on the Crab Nebula since it is such a young, active, and nearby supernova. It has a hard x-ray spectrum, which has been measured by Peterson, Jacobson, Pelling, and Schwartz (1968) in the energy range 20–250 keV, and by Haymes, Ellis, Fishman, Kurfess, and Tucker (1968a) between 40 and 560 MeV. The agreement is reasonably good, and both sets of data are adequately represented by the expression

$$dJ/dE = 5.4E^{-2.1} \text{ photons per cm}^2\text{-sec-keV}.$$

Extrapolation of this spectrum predicts an integral flux, $J(> E)$, per cm²-sec, equal to 1.9×10^{-4} above 10 MeV, 1.5×10^{-5} above 100 MeV, and 1.2×10^{-6} above one GeV. For comparison, Table V lists upper limits that have been reported at various energies.

TABLE V

E	Upper limit of integral flux $(cm^2 \text{ sec})^{-1}$	Reference
> 10 MeV	4×10^{-3}	Fichtel and Kniffen (1965)
> 30 MeV	1.7×10^{-4}	Fichtel et al. (1968)
> 50 MeV	1.7×10^{-5}	Frye and Wang (1969)
> 100 MeV	3.1×10^{-5}	Fazio et al. (1968b)
> 100 MeV	5×10^{-5}	Clark et al. (1968)
> 500 MeV	5×10^{-6}	Frye and Wang (1969)
> 1 GeV	12×10^{-6}	Delvaille et al. (1968)
> 2 TeV	1×10^{-10}	Fazio et al. (1968a)
> 5 TeV	1×10^{-10}	Fruin et al. (1964)
> 5 TeV	5×10^{-11}	Chudakov et al. (1962)
> 1000 TeV	3×10^{-15}	BASJE [Hasegawa et al. (1965)]

A search has also been made for gamma-ray lines in the energy range 50–560 keV by Haymes et al. (1968a) and by A. S. Jacobson (1968). No lines were present with strength as high as 10^{-3} photons/cm²-sec.

One of the most exciting discoveries of the last year was that 5 to 10 percent of the x-ray energy from the Crab Nebula is being emitted in pulses, which are synchronized with the optical and radio pulsations [Fritz, Henry, Meekins, Chubb, and Friedman (1969); Bradt et al. (1969); Fishman, Harnden, and Haymes (1969)].

It would take too long, and not be worthwhile, to list all the negative results in searches for gamma rays from other objects in the sky. Quasars, pulsars, supernovae, x-ray stars, radio galaxies, the Sun and other objects have been examined, with upper limits being found similar to those for the Crab Nebula. Two of these sources deserve special mention, however: the Cygnus x-ray source, Cyg XR-1, and the Virgo A (M87) galaxy.

Cygnus XR-1 is similar to the Crab Nebula, and different from most other sources, in having a hard x-ray spectrum, which has been measured up to 180 keV by Peterson *et al.* (1968), and to 450 keV by Haymes *et al.* (1968b). In fact, the intensity below 100 keV is very similar to that of Tau A in both shape and absolute value. But at about 150 MeV, the spectral index increases from about 1.8 to 2.8, and no gamma rays above 0.5 MeV have yet been detected.

Virgo A is another object with a hard x-ray spectrum. This, however, is not an object in a near arm of our galaxy, but an unusual galaxy at least 10 megaparsecs away. Its x-rays have been measured between 1.2 and 12 keV by Friedman and Byram (1967), between 1.5 and 6 keV by Bradt *et al.* (1967), and between 40 and 100 keV by Haymes *et al.* (1968c). This is the only extragalactic object yet identified as an x-ray or gamma-ray source (with the exception of the Large Magellanic Cloud, a weak but exceptionally near source [H. Mark *et al.* (1969)]). The radio spectrum of M87 is non thermal with a spectral index of -0.75 to -0.79, and the optical radiation from the famous jet is strongly polarized; hence the emission, at least up to the optical region, is synchrotron radiation. It is extremely remarkable that a power-law expression with unchanging exponent, $J(> E) = 40E^{-0.8}$ photons per cm²-sec above energy $E(\mathrm{eV})$, agrees not only with the radio data, but also with the average of the optical flux measurements, and with the x-ray flux from 1 to 100 keV. If this spectrum is extrapolated, it predicts an integral flux above various energies as tabulated in Table VI, in comparison with experimental results.

The comparison at the highest energy indicates that the spectrum certainly steepens below 5 TeV. In view of uncertainties in calibration, however, the negative results of Frye and Wang are not in essential disagreement with the flux predicted by the extrapolation of the radio-to-optical

TABLE VI

Lower energy bound	M 87 Integral flux (cm² sec)⁻¹		Reference
	Extrapolated	Experimental	
50 MeV	2.8×10^{-5}	$<1.5 \times 10^{-5}$	Frye and Wang (1969)
500 MeV	4×10^{-6}	$<1.0 \times 10^{-5}$	Frye and Wang (1969)
1 GeV	2.5×10^{-6}	$(5 \pm 2) 10^{-6}$	Delvaille, Greisen, Albats, and Ögleman (unpublished)
5 TeV	2.8×10^{-9}	$<5 \times 10^{-11}$	Chudakov *et al.* (1962)

spectrum, or with the apparently positive result obtained by the Cornell group (based on a number of counts four standard deviations above background, but still unconvincing and in need of confirmation).

In any case, the spectrum up to 10^5 eV is believable, and represents a radiant flux of 2.6×10^{-9} erg/sec arriving at the Earth. Assuming the source to be at 10 Mpc distance, its luminosity is then at least 3×10^{43} ergs/sec, mostly in the form of x-rays. If the spectrum continues beyond 10^9 eV, the luminosity exceeds 2×10^{44} ergs/sec, mostly in hard gamma rays. For comparison, the total emission of the Crab Nebula is 10^{38} ergs/sec, also primarily in the form of x-rays; and the x-ray luminosity of our Galaxy is estimated to be 10^{40} ergs/sec [Friedman, Byram, and Chubb (1967)].

6.4. Origin of the Isotropic X- and Gamma-Rays

The x-ray flux that is not associated with discrete sources (almost all within the Galaxy) has been shown to be isotropic within about ten percent [Seward *et al.* (1967)], except for the flux at energies below a kilovolt, for which the anisotropy is accounted for by galactic absorption. The spectrum has been discussed in detail above. At energies above 100 MeV, Clark *et al.* (1968) have detected a gamma-ray flux at high galactic latitudes, which also seems to be isotropic, and which matches well with an extrapolation of the isotropic x-ray spectrum. It seems most reasonable to join the points and assume the spectrum to be continuous. The number of photons per unit energy interval goes as $E^{-1.6}$ or $E^{-1.7}$ from one to about 15 keV, and as $E^{-2.1}$ or $E^{-2.2}$ thereafter. An integral of the total arriving energy in quanta above one keV yields 3×10^{-7} erg per cm^2-sec-sr. Below one keV, the spectrum becomes steeper and has only been followed down to 0.27 keV. Mechanisms proposed to account for this isotropic flux are discussed below.

(1)   *Thermal radiation from a dense intergalactic plasma.*

Henry, Fritz, Meekins, Friedman, and Byram (1968) have proposed that the steep spectrum below one kilovolt can be accounted for by free-free emission from an intergalactic gas with a density of 10^{-5} to 10^{-6} atoms per cm^3, at a temperature 3 to 8×10^5 °K. Since $kT = 25$–70 eV, the gas is fully ionized.

Henry *et al.* present convincing arguments that the soft x-rays cannot come from external galaxies in sufficient intensity to account for this flux. The sharp change in spectral index near one keV also gives strong indication that a different emission mechanism is principally responsible below and above this quantum energy. Such information as is available on the spectral *shape* between 0.27 and 1 keV suggests a thermal rather than a non-thermal process. Furthermore, a density of about 10^{-5} atoms per cm^3 is also favored by cosmological theory, and plausible mechanisms for heating the gas are not difficult to find. This proposal, therefore, is quite attractive. More detailed measurements of the spectrum and angular distribution of the soft radiation will ultimately decide its validity. Needless to say, the results will be of great interest to cosmology.

(2)   *Emission from external galaxies.*

Since our Galaxy produces x-rays, it was natural to investigate whether the isotropic flux at energies above one keV could be a super-position of the radiation from similar galaxies distributed through the Universe. Let Q be the emission per galaxy, and n the density of galaxies in the Universe; then the received flux per unit solid angle is $QnL/4\pi$, where L is the effective radius of the Universe. Of course, a cosmologically correct treatment gives a more complicated expression [see, e.g., Brecher and Morrison (1967)]; but for most model universes the simple formula is accurate if one uses for L about $0.5cT$, where T is the age of the Universe. Thus, we take $L = 5 \times 10^{27}$ cm. The galactic density is 10^{-75} per cm^3. As for Q, Friedman, Byram, and Chubb (1967) estimated the emission of our Galaxy at wavelengths between 1 and 10 Å as 7×10^{39}ergs/ sec. Allowing for the x-rays and gamma rays above 12 keV yields a total emission of about 2×10^{40} ergs/sec at energies above one keV. With these values, one finds the expected isotropic flux above one keV to be 8×10^{-9} ergs/cm^2-sec-sr, a factor of 40 below the observed value.

In fact, our Galaxy is a little larger than the average, and if this is also taken into account, the discrepancy is about a factor of 100. Thus, normal galaxies, emitting like our own, cannot account for the isotropic flux above one keV.

Next, it is natural to turn to extraordinary galaxies, such as the intense radio galaxies. However, not only are these galaxies few in number, but apparently they are not exceptionally strong x-ray emitters: the x-rays from Cygnus A, for instance, have not yet been successfully detected. One bright elliptical galaxy, M87, radiates at least 1000 times more strongly than does our Galaxy in x-rays, but M87 is a unique case. Thus, the solution does not seem to lie in this direction.

The above evidence, however, indicates only that the *present* x-ray and gamma-ray emission by normal galaxies and extraordinary radio galaxies is inadequate to account for the background flux. The scattering coefficient of the x and gamma rays is small enough to permit them to have been produced far back in cosmological time, when evolutionary differences in the luminosity and/or density of the sources may outweigh the Doppler effects. In fact, studies of the luminosity distribution of faint radio sources (e.g., Longair, 1966) suggests such an evolution in luminosity or density relative to the coordinate volume, going as a high power (about the cube) of the cosmological scale factor, $1 + z$.

J. Silk (1968) has computed the x-ray background due to normal galaxies under the assumption that the density or luminosity of the galaxies emitting x-rays evolves similarly to the radio sources. Bergamini *et al.* (1968) concluded that the major contribution to the background came from electrons in the strong radio sources (rather than normal galaxies), interacting with the thermal background via inverse Compton effect in the evolutionary past. Maraschi, Perola, and Schwarz (1968) extended the results of Bergamini *et al.* by computing the diffuse gamma radiation arising in this way. In each case, it was found possible to match the observed background flux if one is willing to assume that the evolutionary factor can be applied without a cutoff, as far back in time as $z \cong 10$. The radio observations, however, seem to require a cutoff (presumably representing the time of formation of the galaxies) at a smaller value of z. While this does not preclude a greater evolutionary extrapolation of the x and gamma-ray source functions, it must be admitted that accounting for the isotropic background in this way is highly speculative.

(3)   *Interactions of relativistic particles with the gas, magnetic fields, and radiation fields in intergalactic space.*

(4)   *Interactions of relativistic particles with the gas, magnetic fields, and radiation fields in the halo of the Galaxy.*

The processes in intergalactic space are the same as those occurring in the halo of the Galaxy, so it is efficient to consider both source regions at the same time. Indeed, these processes, occurring in interstellar space in the galactic disk, also give rise to a diffuse but *non*-isotropic flux, which we may as well evaluate simultaneously.

6.5.   Interactions Leading to a Diffuse Flux

(1) *Nuclear collisions $\to \pi^0$ mesons $\to$ gamma rays.*

The gamma-ray spectrum from decay of isotropically moving π^0 mesons has a maximum at $E_m = mc^2/2 = 68\,\text{MeV}$ and if plotted against $\log E$ is symmetric about E_m. Pions of velocity βc produce a flat spectrum of gammas extending from $E_m \sqrt{(1 - \beta)/(1 + \beta)}$ to $E_m \sqrt{(1 + \beta)/(1 - \beta)}$, hence it requires high-energy pions to produce photons of $E \ll E_m$, and

6*

they do it very inefficiently. This process is therefore of negligible consequence in producing x-rays and makes very few gammas below 20 MeV.

The gamma-ray flux is given by

$$dJ(E) = \int J_p(\gamma) \, d\gamma \, d\sigma(E, \gamma) \int n(r) \, dr,$$

where $J_p(\gamma)$ is the cosmic-ray nuclear flux, $d\sigma(E, \gamma)$ is the cross section for generating gammas in dE at E, $n(r)$ is the density of target nuclei, and r is distance from the detector along the line of sight.

The first integral on the right has been evaluated in several publications, using accelerator data for $\sigma(E, \gamma)$ and the cosmic-ray spectrum near the Earth for J_p. Assuming the *shape* of the spectrum to be the same in the source region as at the Earth, the integral gamma spectrum can be written as

$$J_\gamma(>E) = K\bar{n}Lf(E) \, J_p(\text{source})/J_p(\text{Earth}),$$

where K is a constant, the integral of J_p at Earth times the total cross section; L is the depth of the source; n is the average target density; and $f(E)$ is the fraction of the gamma rays having energy above E. Generally n is expressed as the *hydrogen* density in the source (atoms per cm^3), but in the evaluation of K the contributions from heavier nuclei in both the cosmic rays and the interstellar gas are taken into account (mainly p–α and α–p interactions). The publication most extensively quoted prior to 1969 was that of Pollack and Fazio (1963). This gave for the constant K the value 0.97×10^{-26}, of which 2/3 was due to p–p collisions and 1/3 to p–α and α–p events. A very recent evaluation, however (Stecker, 1969), gives for K the value 2.5×10^{-26}, of which only half is due to p–p collisions. The difference is surprisingly large and should be checked. If Stecker's results are applied (with appropriately modified interaction length) to production of gammas in the atmosphere at energies where the geomagnetic effects are negligible, the prediction is too high, whereas good agreement with observations is obtained if the results of Pollack and Fazio are used. Therefore we shall adopt the value $K = 1.0 \times 10^{-26}$. (The difference will not affect our qualitative conclusions.)

For comparison with data on the integral primary gamma-ray flux above various energies, it may be noted that $f(E)$ is about 0.86 at 50 MeV, 0.7 at 100 MeV, and 0.085 at one GeV. At higher energies, $f(E)$ diminishes as $E^{-5/3}$, as inferred from secondary spectra of mesons and photons in the atmosphere.

The product $\bar{n}L$ can be evaluated for the galactic source from neutral hydrogen contours derived from 21-cm radiation (Garmire and

Kraushaar, 1965). Values run from 3×10^{22} cm^{-2} towards the galactic center to about 2×10^{20} towards the pole. On the average at galactic declinations exceeding 40°, one thus predicts a flux of gammas above 100 MeV equal to $10^{-26} (0.7) (3 \times 10^{20}) \cong 2 \times 10^{-6}$ (cm^2-sec-sr)$^{-1}$. This is a factor 50 below the high-latitude flux recorded on OSO-III. Within $\pm 15°$ of the galactic disk, the experiment gave an average flux of 3×10^{-4} (cm^2-sec-sr)$^{-1}$, which was presented as an equivalent line source of 1.6×10^{-4} (cm^2-sec-rad)$^{-1}$. The average value of $\bar{n}L$ in this region of the sky is $(40–50) \times 10^{20}$, giving a predicted flux of 3×10^{-5} (cm^2 sec sr)$^{-1}$, an order of magnitude too low. Towards the center of the Galaxy the experiment gave an average flux of 10^{-3} (cm^2 sec sr)$^{-1}$, while the prediction for an instrument of 15° angular resolution is slightly less than 10^{-4}. In the latter case, the factor J_p (source)/J_p (Earth) may be significantly more than 1 (though most of the line integral of the source strength is in portions of the disk not close to the center), so the existence of a discrepancy is not certain; but this way of adjusting the prediction is not available for the other directions of observation.

Outside the Galaxy both the cosmic-ray flux, J_p, and the gas density, n, are unknown: the hydrogen is ionized and does not reveal itself by 21-cm radiation. However, a generous estimate of J_p is $10^{-2}(J_p$ at Earth) and an upper limit of n is 10^{-5} cm^{-3}, while for L we take 5×10^{27} cm. Thus an upper limit may be calculated for the gamma-ray flux above 100 MeV due to interactions with intergalactic gas. The result is 3×10^{-6} (cm^2 sec sr)$^{-1}$, a factor 30 below the observed isotropic flux.

The conclusion is that the π^0 decay mechanism does not account for the isotropic gamma-ray background, or for the x-ray background, or for the diffuse gamma-ray flux from the disk of the Galaxy.

(2) *Synchrotron radiation.*

Let the differential electron spectrum be $j_e = j_0 \gamma^{-q} d\gamma$ (cm^2 sec sr)$^{-1}$ with γ the Lorentz factor, and let B be the mean magnetic field strength. Then the synchrotron power spectrum is given by the following integral over the line of sight:

$$P(\nu)\, d\nu \cong \frac{2}{9} r_0^2 \left(\frac{e}{2\pi mc} \right)^{(q-3)/2} \nu^{-(q-1)/2}\, d\nu \int j_0 B^{(q+1)/2}\, dr.$$

Translating to number of photons per unit energy, this is

$$J_s(E)\, dE \cong \frac{2}{9} r_0^2 \left(\frac{\hbar e}{mc} \right)^{(q-3)/2} E^{-(q+1)/2}\, dE \int j_0\, B^{(q+1)/2}\, dr.$$

The peak in the radiant spectrum of particular electrons occurs at

$$\nu_m \cong 0.35\gamma^2(Be/2\pi mc).$$

As mentioned in Sec. 5, the radio spectrum in the frequency range 10^7–10^9 Hz agrees well with predictions from these equations, provided (1) the electron spectrum throughout the disk and halo of the Galaxy is the same as that at Earth (with a reasonable correction, $e^{0.4/R(GeV)}$, for solar modulation), throughout the energy interval 0.5–10 GeV, and (2) $B \cong 2\,\mu$G in the halo, $5\,\mu$G in the galactic disk.

Beyond 10^9 Hz there is a limited frequency range in which any continuation of the synchrotron spectrum is obscured by the 2.7° background and other thermal sources. However, if the synchrotron spectrum could be extrapolated to *really* high frequencies it would again become dominant. The question is whether such an extrapolation is a plausible way of accounting for isotropic x and gamma-ray backgrounds.

Assume that at some energy E_c above 10 GeV, the electron spectrum steepens to an exponent of -3.6 because of the radiative energy losses. Then the spectrum of the photons would go as $dE/E^{2.3}$, which is consistent with the diffuse background. However, using $B = 3\,\mu$G and $L = 3 \times 10^{22}$ cm, the above equation for J_s then yields.

$$J_s = 1.8 \times 10^{-8}E_c \text{ (GeV)} E^{-2.3} \text{ (cm}^2 \text{ sec sr MeV)}^{-1}$$

which is low, compared with the observed flux, by a factor 10^5 if $E_c = 10$ GeV. As discussed in the previous section, it is inconceivable that E_c could be higher than this by a very large factor, since that would imply extremely short storage times.

If the radiation actually occurred in fields as weak as $3\,\mu$G, the electron energies required would be 5×10^{14} eV to produce x-rays of 10 keV and 5×10^{16} eV to produce gammas of 100 MeV. Electrons of these energies could not survive long enough even to *reach* the halo. However, since knots of dense field could be responsible for the high-frequency radiation, this argument would not alone have been enough to exclude the synchrotron mechanism.

Similar arguments may be applied to synchrotron radiation in intergalactic space. The available depth L is 10^5 times greater, but E_c is necessarily lower than in the halo by 10^{-2} because of the long storage time; and the weaker fields are responsible (since $J \propto B^{(g+1)/2}$ for reduction of the yield by at least another factor of 10^2. Thus, the synchrotron source is negligible at x-ray or higher quantum energies in both the halo and metagalactic space.

(3) *Compton scattering of fast electrons.*

Felten and Morrison (1963, 1966) are principally responsible for developing the hypothesis that inverse Compton scattering may be responsible for the isotropic x-ray and gamma-ray background. Since the quantum energy after scattering is proportional to the square of the electron's Lorentz factor, the spectrum resulting from the scattering of fast electrons by low-energy photons is closely parallel to the synchrotron spectrum, except for a shift in the frequency scale. That is, the same electrons which produce synchrotron radiation at radiofrequencies in weak fields can produce x-rays or gamma rays by scattering from visible light, infrared, or microwave radiation. Analogous to the expression for J_s in (2) above, we have the following expression for the Compton spectrum:

$$J_C(E)\, dE = \frac{16}{9}\, \pi r_0^2 \left(\frac{4\varepsilon_0}{3} \right)^{(q-3)/2} E^{-(q+1)/2}\, dE \int j_0 u \, dr,$$

where j_0 is the coefficient of the electron spectrum, u is the energy density of the photons and ε_0 is their quantum energy prior to scattering.

It is easy to show that the scattering of visible light and of radio waves in the Galaxy is insufficient to account for the background flux. Scattering of light in intergalactic space (where $u \cong 10^{-2}$ eV/cm^3) could only be adequate if the electron flux were nearly as great in intergalactic space as at the Earth, which is not plausible. However, the 2.7° background radiation discovered in 1965 is much more intense than the optical radiation; its energy density is 0.25 eV/cm^3. Still, the scattering of this radiation by electrons in the Galaxy is also insufficient, and the spectral shape is not quite right, as shown below.

For electrons of energy $0.5 < E < 3$ GeV, we take $J_e = 3.5 \times 10^{-3} E^{-1.6}$ (cm^2 sec sr GeV)$^{-1}$, or $j_0 = 0.33$ and $q = 1.6$. The dimension taken is that of the halo, $\int dr = L \cong 4 \times 10^{22}$ cm; $u = 0.25$ eV/cm^3, and $\varepsilon_0 = 6.3 \times 10^{-4}$ eV. Then the calculated spectrum is $J_C = 0.026 E_\gamma^{-1.3}$ (cm^2 sec sr keV)$^{-1}$ for E_γ in the range 0.8–30 keV. The result of experiments in this energy range was $J_\gamma = 12 E_\gamma^{-1.5}$ in the same units, hundreds of times greater. Similarly, for electrons of $E > 3$ GeV, $J_e = 0.012 E^{-2.6}$ (cm^2 sec sr GeV)$^{-1}$, or $j_0 = 2.2 \times 10^3$ and $q = 2.6$. In this case, the calculation yields $J_C = 0.16 E_\gamma^{-1.8}$ (cm^2 sec sr keV)$^{-1}$, or $6.4 \times 10^{-4} E_\gamma^{-1.8}$ (cm^2 sec sr MeV)$^{-1}$ with E now in MeV units; and this should be valid for $E > 30$ keV. Experimentally $J_\gamma = 1.2 \times 10^{-2} E_\gamma^{-2.3}$ in these units, about 20 times higher at one MeV and differing in exponent by 0.5.

One is then led to consider the same process in intergalactic space. Here the dimension of the source region is 10^5 times that of the galactic halo; hence the computed gamma-ray flux may be adequate even if the electron density is less than that in the Galaxy by a factor varying from several hundred at 1 GeV to 10^4 at 100 GeV. Moreover, the agreement in spectral shape would be better, since energy losses should steepen the electron spectrum by one power of E in intergalactic space, and this would increase the exponent in the calculated gamma-ray spectrum by 0.5.

Compton scattering from the 2.7° radiation in intergalactic space therefore looks like a promising mechanism. But there is still one drawback. If galaxies like ours release their electron content into space with an escape time of 10^6 years from the disk or 10^8 years from the halo, the equilibrium intergalactic electron flux will be less than that in the galaxies by a factor of $10^{-5}/E(\text{GeV})$: too low by more than two orders of magnitude. In fact, in this case the inverse Compton flux from the halo probably exceeds the external flux.

The number of strong radio galaxies is not enough for their electron input into space to provide the needed intensity. And evolutionary effects cannot strengthen the high-energy electron flux in space, since the rate of energy loss is too high. High-energy electrons ($E \gtrsim 1$ GeV) *cannot* have ages close to the age of the Universe.

Cosmology can provide a solution, however, in another way that has already been discussed above, under radiation from external galaxies. Namely, though the electrons cannot survive through cosmological time, the gamma rays can. In an early epoch, radio galaxies were more luminous and/or more numerous than now, and even the space between them was probably more densely populated with fast electrons. Moreover, the background radiation was then at a higher temperature, and its energy density, u, varies as T^4. Therefore the gamma-ray production by Compton scattering probably exceeded the present rate of production by a high enough power of the expansion factor, $1 + z$, to overcome strongly the Doppler factors associated with the universal expansion. I.e., it is likely that the x-ray and gamm a-ray background is a relic of an early phase of the Universe, when most of the galaxies were being formed.

(4) *Effects of the possible infrared background.*

The apparent discovery of an intense background by Shivanandan, Houck, and Harwit (1968) provides an alternative to the cosmological explanation of the background radiation. Again the basic process is Compton scattering of the fast electrons. However, the infrared detected by these a uthors has an apparent energy density at least fifty times that of the 2.7°

radiation, so the problems of obtaining an adequate scattered intensity are largely removed.

The explanation of the diffuse radiation on this basis has been put forward by Shen (1969) and by Cowsik and Pal (1969).

If the IR flux is indeed universal, the electron energy loss rate is so great that intergalactic production of gamma rays is negligible. The production therefore occurs in the disk and halo of the Galaxy, and is predicted (by equations very similar to those applied above to scattering by the 2.7° radiation) to be of the right order of magnitude in intensity. Moreover, this model provides in a natural way for the excess flux observed in the plane of the Galaxy, and particularly towards its center, by the OSO-III experiment at 100 MeV. However, the following criticisms of the model can be made:

(a) It has not yet been proved that the diffuse IR detected by Shivanandan *et al.* is universal rather than a local phenomenon.

(b) Observations of interstellar optical absorption lines corresponding to rotational states of CN, CH, and CH+ (Bortolot, Clauser, and Thaddeus, 1969) agree with the existence of background radiation at about 3°K, but are inconsistent with the existence of the large amount of IR, unless it is fortuitously concentrated in lines that escape the molecular resonances.

(c) As pointed out in Sec. 5, the flatness of the electron spectrum at very high energy (100 GeV) argues against the existence of such intense IR throughout the Galaxy. The radio flux from the halo indicates no break in the electron spectrum in the halo up to nearly 10 GeV, and this also argues strongly against universality of the IR.

(d) The flatness of the primary proton spectrum between 10^{19} and 10^{20} eV is also hard to account for if the IR is universal.

(e) The exponent of the predicted Compton recoil spectrum does not agree well with observations: the prediction is $E^{-1.8}$ while the observed spectrum follows more nearly $E^{-2.2}$.

(f) The excess flux in the galactic plane is very likely not to be a diffuse flux but the superposition of radiation from discrete objects, like the x-ray flux (Ögelman, 1969).

(g) As pointed out by O'Connell and Verma (1969), if the scattering yields a diffuse gamma-ray background, it necessarily produces a similar, predictable x-ray background. However, owing to variation in the thickness of the Galaxy along various lines of sight from the Earth, this background will be far from isotropic, varying by a factor of 5 from maximum to minimum outside the galactic disk. The gamma-ray diffuse background is not yet known to be accurately isotropic, but the x-ray background is isotropic to within 10%.

These arguments seem to me to be very strong ones. Clearly the question of the infrared intensity in space is crucial and must be settled by further observations, but for the present the cosmological explanation of the origin of the isotropic diffuse x-ray and gamma-ray flux seems more consistent with existing information. Along with this conclusion, I favor the model of discrete sources to account for most of the excess gamma-rays (and x-rays) near the galactic plane.

(5) Bremsstrahlung.

Bremsstrahlung is an efficient process by which electrons inter-act with the ambient gas to generate gamma rays of all energies up to that of the electrons themselves. In first approximation, the gamma-ray flux is given by

$$dJ_\gamma(E) = \frac{dE}{E} \frac{M(L)}{X_0} J_e(>E),$$

where $J_e(>E)$ is the *integral* electron flux above energy E, X_0 is the radiation length (63 g/cm² in hydrogen) and $M(L)$ is the integrated density of the gas in g/cm² along the line of sight. If the electron spectrum is representable by a power law for electron energies above the gamma-ray energy in question, the electron and bremsstrahlung gamma-ray spectra are parallel, and a more accurate expression is

$$dJ_\gamma = dJ_e \cdot M(L) X_0^{-1} f(q),$$

where $f(q)$ is a numerical factor near one, depending on the exponent of the differential energy spectra, q, and on the screening correction. For $E \ll 70$ MeV, $f(q) = [1/(q-1)](4/3q)$, and for $E \gg 70$ MeV, $f(q) = [1/(q-1)](1 + 1/3q)$.

Values of $M(L)$ for directions of special interest are given below.

Direction	$M(L)(g/cm^2)$
Galactic center	6×10^{-2}
Anticenter	1.2×10^{-2}
Pole of Galaxy	6×10^{-4}
Average over Galaxy	1.6×10^{-3}

At high energies, above 3 GeV (see Sec. 5) the electron flux is given by $dJ_e = 1.2 \times 10^{-2} E^{-2.6} dE$ per cm²-sec-sr, with E in GeV. The bremsstrahlung photon spectrum at high galactic latitudes is therefore expected to be

$$dJ_B = 8 \times 10^{-8}E^{-2.6} \, dE \text{ per cm}^2\text{-sec-sr}.$$

This may be compared with the expected yield from pion decay, which is $4 \times 10^{-7}E^{-8/3} \, dE$: five times the yield from bremsstrahlung. The ratio is independent of errors in assumed mean gas density, and is the same in other directions through the Galaxy. (The difference between the two exponents is negligible, and less than their errors.)

At energies between 50 MeV and 3 GeV the photon spectra arising from both of these processes are less steep than at high energy, but remain roughly parallel and the pion mechanism continues to dominate. Since the latter was inadequate to account for the observed gamma-ray background, the bremsstrahlung yield is negligible. At still lower energies, since the bremsstrahlung process has no threshold, the yield by this process surpasses that from pion decay; but it is less than the observed gamma-ray flux around one MeV by a factor of about 10^4.

Thus, the bremsstrahlung process in interstellar space is completely inconsequential in accounting for the diffuse x and gamma rays.

Summary

The soft x-ray background below 1 keV may be accounted for by thermal radiation from a hot intergalactic plasma: the spectrum is not yet known in sufficient detail to confirm or negate this hypothesis. But neither the isotropic x and gamma-ray background, nor the excess gamma rays from the galactic plane, are satisfactorily explained. Pion production and decay is insufficient, and cannot be increased by postulating more gas and/or cosmic rays in space without predicting more positrons than are found. Bremsstrahlung is even more inadequate, and synchrotron radiation is weaker still. Inverse Compton scattering of the 2.7° background radiation by electrons in the Galaxy is inadequate by one to two orders of magnitude and does not yield the right spectral shape. The same scattering process, occurring in intergalactic space, can only supply the observed radiation if the electron flux is assumed to be quite implausibly high: one does not solve anything in this way, but merely shifts the dilemma from accounting for gamma rays to explaining so many electrons. Inverse Compton scattering from an intense infrared flux, recently observed at rocket altitudes, might be sufficient if the IR flux is universal; but there is strong evidence against this, and also the predicted spectral shape would not be in good agreement with the data.

Having found all sources in the present Universe inadequate, we are left with the remote past. It seems likely that the background x and gamma radiation is a relic of intense production within and between the galaxies in an earlier epoch when the galaxies were being formed, and the number and luminosity of strong radio sources were much greater than at present. The temperature and energy density of the background radiation (now 2.7°) were also much higher at that time; and the responsible process might well have been inverse Compton scattering. At the epoch characterized by the expansion factor $1 + z$, the radiant energy density was higher than now by $(1 + z)^4$ and the average electron flux may have been higher by a factor $(1 + z)^3$. The fact that this source was receding from us cancels the first of these factors but not the second, so the residual density of the radiation may be much greater than that of the radiation currently being produced. However, quantitative analyses of gamma-ray production in the era of galaxy formation presently require many unsupported assumptions to fill in essential gaps in our knowledge, and cannot yet be regarded with confidence.

The following test may be suggested, however. The density of optical radiation in the epoch of gamma-ray production would have been great enough to absorb the gamma rays above 10^{11} eV by the process of pair production ($\gamma + \gamma \rightarrow e^+ + e^-$). The subsequent expansion would have shifted the cutoff downwards in energy by a factor $1 + z$. Therefore we predict that if the cosmological origin of the background gamma radiation is correct, the present spectrum should have an abrupt cutoff in the neighborhood of 10 GeV.

6.6. Origin of Gamma-Rays in Discrete Sources

The processes are the same ones that have been discussed above in this section and Sec. 4; here we attempt to evaluate the expected gamma-ray flux at the Earth and compare it with the upper limits that have been determined. (No positive measurement of gamma-ray flux from discrete sources at energy above one MeV has yet been accomplished). Particular application will be made to the Crab Nebula as an example, rather than constructing a catalogue of expectations for numerous cases; but the dependence of the expected flux on source characteristics will be shown, so as to facilitate applications to other instances.

(1) *Gas collisions* ($p + p \rightarrow \pi^0 \rightarrow \gamma + \gamma$, *mostly above* 50 MeV).

$$J_\gamma \cong N_p \frac{c}{4\pi} \frac{n\sigma}{R^2},$$

where J_γ is the total gamma-ray flux at the Earth, N_p is the total number of relativistic protons in the source, n is the number per unit volume of target atoms in the source (which is necessarily thin), σ is the cross section times average multiplicity of gamma rays, and R is the distance from the Earth. We take $R = 1700$ pc ($4\pi R^2 = 3.4 \times 10^{44}$ cm^2), $\sigma = 3 \times 10^{-26}$ cm^2, and $n \cong 10$ cm^{-3} (about one solar mass in the nebula). N_p is in more doubt. There are about 10^{50} electrons above 100 MeV in the nebula (assuming $B = 5 \times 10^{-4}$ gauss and $R = 1700$ pc), and the usual ratio of protons to electrons is 100, so let us try $N_p = 10^{52}$. The result is then $J_\gamma = 2.6 \times 10^{-7}$ (cm^2 sec)$^{-1}$ at Earth, well below the upper limits set by experiments.

Alternatively, it was once popular to set the proton spectrum to be whatever is necessary to *produce* the electrons that are in the nebula in its lifetime or (for high energies) in the half-life for electron energy loss. The calculation is straightforward and many factors, such as the nuclear cross section and the distance to the nebula, cancel out. We omit the details. The results at a few energies are:

$$J_\gamma(>10^8 \text{ eV}) \quad = 1.2 \times 10^{-5} \text{ (cm}^2\text{-sec)}^{-1},$$

$$J_\gamma(>250 \text{ GeV}) = 3 \times 10^{-8} \text{ (cm}^2\text{-sec)}^{-1},$$

$$J_\gamma(>5 \text{ TeV}) \quad = 5 \times 10^{-10} \text{ (cm}^2\text{-sec)}^{-1}.$$

This model for the associated origin of the electrons and gamma rays can be ruled out on several grounds. First, if high-energy electrons originated primarily by nuclear collisions, we would have more positrons than electrons in the cosmic rays, instead of the reverse. Secondly, the gamma-ray flux predicted at the two highest energies listed above is much more than the upper limits set by the Čerenkov shower-detecting experiments, discussed earlier in this section. And thirdly, such a large number of high-energy protons would have generated an outwards acceleration of the gas that has not occurred. On the latter basis, even $N_p = 10^{52}$ is somewhat too high unless the mass of the nebula is several times that of the Sun.

(2) *Thermal radiation.*

For some of the x-ray sources, the observed spectra have the shape expected for thermal radiation from a thin, hot, plasma, namely $dJ \sim e^{-h\nu/kT} d\nu/\nu$, with kT varying from about 0.5 keV for the Sun to 5 keV for sources like Scorpius X-1. However, this shape is so steep at energies $\gg kT$ that its extension to gamma-ray energies is unobservable.

(3) *Radioactivity.*

In supernova explosions, the R-process (rapid neutron-reaction buildup of heavy elements) is expected to yield radioactive elements of long life. These should reveal themselves by lines in the spectrum at energies below 0.5 MeV. However, the predicted flux in the strongest cases is about 10^{-5} photons per cm²-sec, while the experimental limits of detectability have been much higher, about 10^{-3} per cm²-sec. There are several sources from which hard x-rays have been detected up to 0.5 MeV, but in all these cases a continuous spectrum is dominant, obscuring any lines that may be present.

(4) *Bremsstrahlung.*

Representing the electrons in the source by $dN = N_0 E^{-q} dE$, the bremsstrahlung flux at the Earth can be estimated by

$$dJ_\gamma = \frac{f(q)}{q-1} \frac{c}{4\pi R^2} \frac{\bar{n}M}{X_0} N_0 E^{-q} dE,$$

where M is the atomic mass of the gas atoms and the other symbols have been defined above. For the Crab Nebula, for instance, if we take $\bar{n} = 10$ atoms of hydrogen per cm³, $R = 1700$ pc, and the number of electrons above 100 MeV as 8×10^{49}, we find that the number of gamma rays above 100 MeV is expected to be 4×10^{-9} per cm²-sec, with a spectrum going about as $E^{-1.6} dE$. Such a flux is far below the limits of detectability, and despite the uncertainties in the estimates, is clearly less than the flux expected from π^0 decay.

(5) *Synchrotron radiation (magnetic bremsstrahlung).*

For an electron spectrum given by $dN = N_0 E^{-q} dE$ (with E expressed in ergs) with the electrons moving isotropically in a magnetic field of B gauss, at a distance R from the Earth, the synchrotron power spectrum (in ergs per cm²-sec-Hz) is given by Ginzburg and Syrovatskii (1964) as

$$I(\nu) = 1.35 \times 10^{-22} a(q) \frac{N_0}{R^2} B^{(q+1)/2} \left(\frac{6.26 \times 10^{18}}{\nu} \right)^{(q-1)/2},$$

where $a(q)$ is a slowly varying numerical factor: 0.15 for $q = 1.5$; 0.074 for $q = 3$.

For a source like the Crab Nebula, the radiation in the radio-frequency and optical range is identified as synchrotron radiation by its polarization and spectral shape. We do not know the electron spectrum independently, but infer it from this equation. See Sec. 3 for a discussion of how one estimates the value of the field strength B. This parameter, as well as the distance R, is in some doubt: the measurements determine directly only the quantity $N_0 R^{-2} B^{(q+1)/2}$.

The radiation from the Crab Nebula continues as a smooth power law beyond the visible, up to quantum energies of at least 0.5 MeV. The continuity of the spectrum tempts one to assume that synchrotron radiation remains the responsible process. However, the necessary electron energy rises impressively. To emit radio waves of 10^8 Hz, one needs electrons of $E \cong 300$ MeV; for the visible blue light one needs $E \cong 10^{12}$ eV; for x-rays of 40 keV, $E \cong 10^{14}$ eV, and for photons of 0.4 MeV, $E \cong 3 \times 10^{14}$ eV. It does not seem likely that the process can continue much further, if it continues even this far.

In any case, the electron spectrum in the Crab Nebula is much steeper beyond 10^{12} eV than below 10^{11} eV, as revealed by a change in slope of the radiation spectrum in the interval 0.1–1 eV of quantum energy. We shall see below that there are other mechanisms which, on this account, should assume dominance in producing high-energy photons.

(6) *Bootstrapping: the Compton–synchrotron process.*

The synchrotron radiation remains in the nebula for a time on the order of r/c where r is the radius of the source. Within the Crab Nebula, for instance, this radiation has an energy density of about 20 eV per cm³, mostly in quanta of 0.1 to 10 eV energy. The electrons which produced this radiation can undergo Compton scattering from it, producing outgoing radiation of quantum energy γ^2 (where γ = Lorentz factor) times that of the synchrotron radiation. By this two-step radiative process, the electrons can generate photons of energy up to nearly that of the electrons themselves.

Let the intensity of synchrotron radiation reaching the Earth be $I(\nu)$, and assume that this comes from a spherical nebula of radius r at a distance R from the Earth. Then the number density of the synchrotron radiation in the nebula as a function of quantum energy, $\varepsilon = h\nu$, is

$$d\varrho(\varepsilon) = \frac{12}{h^2 c \theta^2} \frac{I(\nu)}{\nu} d\varepsilon,$$

where $\theta = 2r/R$ is the angular diameter of the nebula. We will treat the simple case where the synchrotron radiation has a power spectrum $I(\nu) = I_0(\nu/\nu_0)^{-\alpha}$, and arises from relativistic electrons, whose number as a function of Lorentz factor is $N(\gamma)$. As noted in the previous section, $N(\gamma)$ is inferred from $I(\nu)$ and has the form $N_0 \gamma^{-q}$, with $q = 1 + 2\alpha$ and N_0 proportional to $I_0 R^2 B^{-(\alpha+1)}$. We further assume, for the present, that $\varepsilon\gamma \ll mc^2$, so that the Compton cross section is a constant, $\sigma_0 = 0.66 \times 10^{-24}$ cm², and the scattering is symmetric in the electron rest-frame. Averaging over scattering angle, the energy of the scattered radiation is related to ε and γ by $E = (4/3)\,\gamma^2\varepsilon$, and the flux at the Earth is

$$dJ_{C-s}(E) = \frac{c\sigma_0}{4\pi R^2}\int N(\gamma)\,d\gamma \cdot d\varrho(\varepsilon)$$

Therefore,

$$dJ_{C-s}(E) \propto \frac{I_0^2\sigma_0}{\theta^2 B^{\alpha+1}}\int \frac{d\varepsilon}{\varepsilon^{\alpha+1}}\,\frac{d\gamma}{\gamma^q}$$

$$\propto \frac{I_0^2\sigma_0}{\theta^2 B^{\alpha+1}}\,\frac{dE}{E^{\alpha+1}}\int \frac{d\gamma}{\gamma}.$$

It is apparent that a finite result is only obtained if the assumed spectra are limited in extent. For instance, we may consider the portion of the synchrotron spectrum of the Crab Nebula at frequencies below $\nu_1 \cong 10^{14}$ Hz, which corresponds to a power law with $\alpha = 0.27$, and is due to an electron spectrum extending up to $\gamma_1 \cong (6\pi mc\nu_1/Be)^{1/2}$ with exponent $q = 2\alpha + 1 = 1.54$. The maximum scattered photon energy, $E_{\max}$, is then $(4/3)\,h\nu_1\gamma_1^2 \cong 6\times 10^7 B^{-1}$ eV; and for values of E below this, the minimum value of γ is $\gamma_{\min} = (3E/4h\nu_1)^{1/2}$. The integral $\int d\gamma/\gamma$ then has the value $\frac{1}{2}\ln(E_{\max}/E)$ and we have

$$dJ_{C-s}(E) \propto \frac{I_0^2\sigma_0}{\theta^2 B^{\alpha+1}}\,\frac{dE}{E^{\alpha+1}}\ln\frac{E_{\max}}{E}$$

$$\text{for}\quad E < E_{\max} = \frac{8\pi hmc}{Be}\,\nu_1^2.$$

This is not the entire result. At frequencies above ν_1, the synchrotron spectrum continues at a steeper slope, $\alpha \cong 1.1$, corresponding to a high-energy branch of the electron spectrum with $q \cong 3.2$. One should take into account the interaction of each branch of the electron spectrum with each branch of the synchrotron spectrum. Moreover, in the interactions leading to very high E values the condition $\gamma\varepsilon \ll mc^2$ fails, and one must use the accurate Compton cross section. Instead of giving details here, we refer to the original paper on this subject by R. J. Gould (1965) and the more recent calculations published by Rieke and Weekes (1969). The principal effect of the more inclusive treatment is to add an extension of the Compton spectrum at energies above the previous $E_{\max}$. The extended spectrum is quite steep, going about as dE/E^3, in contrast to $dE/E^{1.3}$ below $E_{\max}$.

The relation between the synchrotron spectrum and the Compton-scattered radiation is pictured in the accompanying qualitative graph (Fig. 4). The ratio of total energies in these two spectra is the ratio of magnetic-field energy to the radiant energy in the nebula, which is about 100 to 1. The Compton radiation is completely negligible in the x-ray and lower-frequency regions. It achieves prominence at higher frequencies only because of the steepening or cutoff of the electron and synchrotron spectra, but this permits it to dominate at energies on the order of 10^8 eV and higher. The integral flux at Earth as calculated by Gould (1965) was about 10^{-6} photons per cm²-sec above 10^8 eV, 5×10^{-7} above 10^9 eV and 2×10^{-7} above 10^{10} eV, assuming an average field strength of 10^{-4} gauss in the nebula.

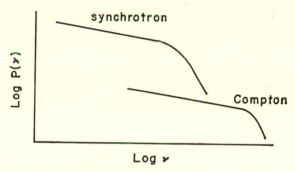

FIG. 4. Photon spectra generated via the synchrotron and inverse Compton processes, by the same spectrum of electrons. The relative displacement of the two photon spectra depends on the field strength and angular diameter of the source, as indicated in the ext.

The main uncertainty in these predictions arises from lack of knowledge of the field strength. The dependence on B is displayed in the equations above ($J \propto B^{-(\alpha+1)}$), and arises because the number of electrons is inferred from the synchrotron radiation. Note also that the energy at which the Compton spectrum steepens is inversely proportional to B. Thus, if B is taken to be 10^{-3} gauss instead of 10^{-4} as assumed by Gould, the flux values are smaller by a factor of about 20, and the abrupt steepening occurs at 10^{10} instead of 10^{11} eV.

Because of these relationships, if one can succeed in measuring the inverse Compton radiation, it will constitute a determination of the average field strength in the source. At present, experiments have given only upper limits of the high-energy gamma radiation from the Crab Nebula and other sources. These results imply lower limits of the field strength. Thus, it has already been concluded on this basis (Fazio, Helmken, Rieke, and Weekes, 1969) that the average field strength in the Crab Nebula cannot be less than 10^{-4} gauss.

(7) *Compton scattering of external radiation.*

It was pointed out by Apparao (1967) that the electrons in a nebula can scatter not only from their synchrotron radiation, but also from the 2.7° background that illuminates all space (and also from the infrared radiation discovered by Shivanandan, Houck, and Harwit, if this proves to be universal). In this way, the 2.7° radiation acts like a searchlight, picking out places in the Universe that contain high-energy electrons, and revealing their number and energy by the intensity and quantum energy of the reflected gamma rays.

Let ϱ_0 be the number density of the 2.7° radiation, and assume all the photons to have the mean energy $\varepsilon_0 = 2.7kT = 6.3 \times 10^{-4}$ eV. The scattered flux from the electrons in a source at distance R can then be expressed by

$$dJ(E) = \frac{\varrho_0 c \sigma_0}{4\pi R^2} N(\gamma) \, d\gamma \quad \text{with} \quad E = \frac{4}{3}\gamma^2 \varepsilon_0$$

and
$$N = N_0 \gamma^{-q}.$$

Thus,

$$dJ(E) = \frac{c\sigma_0}{8\pi} \left(\frac{4}{3}\right)^\alpha \frac{N_0}{R^2} \varrho_0 \varepsilon_0^\alpha \frac{dE}{E^{\alpha+1}},$$

where $\alpha = (q - 1)/2$ is the power index of the synchrotron radiation from the same electrons. As before, N_0 is inferred from the synchrotron radiation, and is proportional to $I_0 R^2 B^{-(\alpha+1)}$, I_0 being the coefficient of the synchrotron power spectrum. Therefore,

$$dJ(E) \propto \frac{I_0 \sigma_0}{B^{\alpha+1}} \varrho_0 \varepsilon_0^\alpha \frac{dE}{E^{\alpha+1}}.$$

This form continues up to an energy E determined either by a maximum value of γ in the electron spectrum, or a breakdown of the condition $\gamma\varepsilon_0 \ll mc^2$, which is equivalent to $E \ll 4(mc^2)^2(3\varepsilon_0) \cong 5 \times 10^{14}$ eV. In the case of the Crab Nebula, the electron spectrum steepens at a Lorentz factor γ_1, leading to a sharp steepening of the scattered radiation spectrum beyond an energy of $10^5 B^{-1}$ eV.

One notices immediately that the spectrum is parallel to the Compton–synchrotron spectrum, and it is of interest to compare their intensities. From the relations given above one can see that

$$\frac{J_{C-s}}{J_{C-3^\circ}} \propto \frac{I_0}{\theta^2},$$

i.e., to the received synchrotron radiation per unit solid angle. Scattering of the background microwave radiation is apt to dominate from expanded, diffuse sources, and Compton–synchrotron radiation will dominate from compact sources: especially from quasars.

At energies below the points of steepening of the scattered spectra, the two processes yield approximately equal numbers of gamma rays in the case of the Crab Nebula. This seems surprising at first, since the energy density of the 2.7° background is only 0.25 eV/cm³, about 80 times less than the energy density of synchrotron radiation in the nebula. The reason becomes apparent upon examining the above relations for $dJ(E)$ and observing that each photon is weighted with its energy to the power, α, and in the case of the Crab Nebula, α is remarkably small, about 0.27. Most of the energy in the Crab's synchrotron radiation is carried by quanta with frequency near ν_1, about a thousand times more energetic than the 3° radiation. The relative yield predicted below $10^5 B^{-1}$ eV, is

$$\frac{J_{C-s}}{J_{3^\circ}} = (2 - \alpha) \frac{u_s}{u_{3^\circ}} \left(\frac{\varepsilon_0}{\varepsilon_1}\right)^{1-\alpha} \cong 1.1.$$

In sources with steeper electron spectra, the Compton–synchrotron process will have greater prominance relative to the scattering of background radiation.

We cannot end this section, however, without a reminder of the great uncertainty in the amount of infrared radiation in the Universe. The tentative evidence, discussed repeatedly above, is for an isotropic infrared component with energy density two orders of magnitude more than that of the 2.7° background. If this should prove real, the Compton scattering of the infrared would be the dominant mechanism of gamma-ray yield from many sources, and the predicted intensity would be very close to the limits that have already been set experimentally.

Summary

In contrast to the difficulty of finding adeqate mechanisms to account for the isotropic gamma radiation which has actually been detected, there are numerous mechanism for production of the gamma rays we have *not* been able to detect from discrete sources. No one of these mechanisms is dominant under all circumstances. Thermal radiation produces the biggest fluxes, but only at soft x-ray frequencies. Synchrotron radiation dominates at very low (radio) frequencies and in some cases, apparently all the way up to the hard x-ray range or further. Pion decay resulting from gas collisions competes with inverse Compton scattering, both of synchrotron radiation and of external radiation, in providing high-energy quanta, above 10^8 eV. The expected variations in density, age, and means of excitation in various potential sources will probably make each of these mechanisms dominant in some instances. Already the experimental flux limits have determined significant minima for the field strength in some sources. and improvement in detector sensitivity by one more order of magnitude is very likely to result in the positive detection of a number of sources and determination of their emission mechanisms.

7. COSMIC X-RAYS

7.1.   Foreword

A large part of the subject of cosmic x-rays has already been presented in prior sections, because of its intimate relation to other components of cosmic rays and their processes of emission. The diffuse x-ray background, in particular, has been thoroughly discussed in Sec. 6. In the present section, therefore, I will present a rather condensed summary of what is known about cosmic x-rays, with particular emphasis on discrete sources and very little further mention of the isotropic background.

Of particular value to the student of this subject are the survey article, "Origin of Cosmic X-Rays" by R. J. Gould (1967), and the recent review, "Discrete Extrasolar X-Ray Sources" by B. Rossi (1968). Observational techniques have been summarized by Giacconi, Gursky, and van Speybroeck (1968). A careful theoretical analysis of models of the x-ray sources has been given in the article, "Cosmic X-Ray Sources", by Wallace Tucker (1967).

7.2.   Introduction

The discovery of nonsolar x-ray sources in 1962 was quite unexpected—in contrast to high-energy gamma-ray sources, which were predicted much earlier, but have not yet been observed. In some measure, this error was due to being overly impressed by the dominance of nuclei over electrons in the cosmic rays, and therefore assuming the electrons to be a secondary component generated by pion production. Pion decay would have yielded abundant gamma rays above 50 MeV, but practically no x-rays. As explained in earlier sections, these expectations were wrong: the electrons are mostly not secondaries, and nuclear collisions are apparently not the major source of the electromagnetic radiation—even the high-energy gamma rays. A second impediment was the conditioning of thought by what was known of x-rays from the Sun. Although greater yields were predicted for other objects, including peculiar stars and supernovae, the distances to such sources are so large that the anticipated flux at the Earth was thousands of times less than that from the Sun, and would have been very hard to detect indeed.

Solar x-ray emission had been known for more than a decade and had been studied especially by scientists under H. Friedman at the Naval Research Laboratory. The spectrum contains strong lines at long wavelengths: e.g., carbon K lines at about 25 Å and oxygen at 14 Å. Below 10 Å (i.e., at energies above one keV), there is a continuum resembling the emission of a hot plasma at $T = 4.5 \times 10^6$ °K, or $kT = 400$ eV (Chodil et al., 1965). Although this flux contains thousands of x-ray photons per cm^2-sec, the power spectrum is exponential according to $e^{-h\nu/kT}$, and is undetectably weak above about 5 kilovolts. The average total energy flux in x-rays above one kilovolt is only 10^{-10} of the solar luminosity in the visible and infrared.

Prior to 1962, scientists at MIT and ASE (American Science and Engineering) prepared an x-ray detector for a rocket flight in an effort to learn about the Moon by detecting anticipated fluorescent x-rays resulting from solar radiation. The instrument had very crude angular resolution (60°), but as the rocket spun at heights up to 230 km, a modulated x-ray signal was detected, coming not from the Sun or the Moon, but from a direction in Scorpius, near (but slightly displaced from) the center of the Galaxy. At quantum energies above 4 keV, this source was stronger than the Sun and dominated the whole sky. Thus were nonsolar x-rays discovered by Giacconi, Gursky, Paolini, and Rossi (1962), and a new era of astronomy was begun.

One should try to appreciate how amazing this discovery was at the time, in spite of the fact that we have come to accept it as a matter of course by now. The energy flux above one kilovolt from Scorpius X-1 is 5×10^{-7} ergs/cm^2-sec. The distance was not known at the time, though it was clear that it was at least many tens of light-years. More recently, this source has been optically identified (Gursky et al., 1966a; Sandage et al., 1966) and the optical interstellar CaII absorption linewidths have been found to imply a distance, R, exceeding 270 pc (Wallerstein, 1967). The indetectability of proper motion suggests a distance several times this value; but since the source is in the galactic disk, its rather high galactic latitude (24°) implies that R probably does not exceed 500 pc. Thus R is about 10^{21} cm or more, implying $4\pi R^2 \geq 10^{43}$ cm^2 and that the x-ray luminosity is at least 5×10^{36} ergs/sec, a thousand times the *total* luminosity of the Sun, and 10^{13} times the x-ray luminosity of the Sun.

The star with which Sco X-1 has been identified is rather faint: 13th magnitude. Therefore the x-ray luminosity is a thousand times its optical luminosity, while the solar x-ray emission is 10^{-10} times its optical emission. In terms of relative luminosity in the x-ray and visible parts of the spectrum, Sco X-1 differs from the Sun by a factor 10^{13}.

Not only is the optical emission comparatively weak from Sco X-1, but its radio emission is too faint to detect. Truly this was an example of a new class of objects, an x-ray star.

7.3.   Identifications of X-Ray Sources

Since 1962 about forty other x-ray sources have been detected—the exact number being unknown, since sources observed by different experimenters at nearly the same angular coordinates are in some cases, but not all, the same objects. But of all these x-ray sources, only a small number have been identified with optical counterparts: four or five cases with virtual certainty, and four or five others that must still be regarded as tentative. They are listed below in Table VII with brief comments, after which a more detailed discussion of a few of the associations will be given.

The first of the above identifications to be made was that of the Taurus x-ray source with the Crab Nebula. Suspecting the association on the basis of approximately known coordinates of the x-ray source, the NRL group (Bowyer *et al.*, 1964) took advantage of a rare opportunity, an impending eclipse of the nebula by the Moon, and timed a rocket flight to embrace the period of the eclipse. Not only did the observation confirm the association, but it located the position of the x-ray source within the nebula, and also showed that the source was diffuse, one or two minutes of arc in diameter.

(Interestingly, this observation was taken to prove that the x-ray source was not a neutron star, since that would make it appear to be a point source. Now, five years later, it has been discovered that this obvious conclusion was too simple. In fact, the source is compound: most of the

TABLE VII

Source	Reference	Remarks
(1) Tau A	Bowyer *et al.* (1964) Oda *et al.* (1967)	Supernova 1054; Crab Nebula. Positive
(2) Cas A	Byram *et al.* (1966) Friedman *et al.* (1967)	SN II (1720). Faint; identification tentative
(3) Cas B	Friedman *et al.* (1967)	Tycho SN I (1572). Faint; tentative
(4) Vir A (M 87; NGC 4486)	Byram *et al.* (1966); Friedman and Byram (1967); Bradt *et al.* (1967); Haymes *et al.* (1968)	Radiogalaxy ("Jet" galaxy at $d \cong 11$ Mpc); identif. strong
(5) GX 3 + 1	Bradt *et al.* (1968) Blanco *et al.* (1968) Goss *et al.* (1968)	Tentative ident. with supernova (or extreme Wolf-Rayet star)
(6) Sco X — 1	Gursky *et al.* (1966*a, b*) Sandage *et al.* (1966)	Very strong ident.; variable blue star of mag 12.5
(7) Cyg X — 2	Giacconi *et al.* (1967*a*)	Strong ident.; variable blue star of mag 15
(8) Cen XR — 2	Eggen, Freeman, and Sandage (1968)	Tentative ident. with WX Cen, variable blue star of mag 13–14
(9) Large Magellanic Cloud	Mark *et al.* (1969)	Source 12° wide; a super-position of all x-ray sources within the LMC

x-rays indeed come from a broad region, but about 9% come from a point-like, pulsing object at its center, NP 0532, which is apparently a neutron star [Fritz *et al.* (1969); Bradt *et al.* (1969).]

The position and size of the x-ray source have also been determined with remarkable precision, using the technique of fine wire modulation collimators, by the MIT-ASE group (Oda *et al.*, 1967). The position is well centered within the nebula (centered on NP 0532), and the diameter is about 100 arc seconds, implying a linear diameter of 2.7 light-years if the distance is 1700 pc (5500 lt-yr).

The next identification, that of Sco X-1, was harder because there seemed to be no unique candidate among the many optical sources in about the right direction. With a modulation-grid collimator, Gursky *et al.* (1966*b*) determined an upper limit of 20 arc seconds (10^{-4} rad) to the diameter of Sco X-1. Assuming the source to be starlike and the emission to be by thermal bremsstrahlung, the spectrum was extrapolated from x-ray frequencies to the visible and a prediction was made that the object would be of 13th magnitude with specific spectral characteristics (strong ultraviolet excess). With the same modulation collimator (Gursky *et al.*, 1966*a*), the source was determined to lie within one or another of two small regions of the sky, each 1 × 2 arc minutes in size (or with lower probability, in one of two further regions of like size). The coordinates were communicated to observatories in Japan and California, where it was found that just one optical candidate existed that met the specifications (Sandage *et al.*, 1966); and this lay in one of the two more probable boxes.

Subsequent identifications have been facilitated by these first successes. Within the error radius of the x-ray source coordinates, usually much larger than one arc minute, it is not unlikely to find a number of optical sources. But if the search is restricted to sources having very rare characteristics, the likelihood of false identification occurring by chance is small. In the case of Cygnus X-2, the error radius was about ten arc minutes (Giacconi *et al.*, 1967*b*); but there was only one candidate having both the right magnitude and spectral character. Moreover, like the optical counterpart of Sco X-1, this star was an optical variable.

In the case of Cen X-2, the x-ray error radius was 1.5°, but the candidate star (WX Cen) was found to have many spectral features like the unusual ones of Sco X-1, including the *B–V* and *U–B* indices, many broad emission lines, and strong variability (Eggen, Freeman, and Sandage, 1968).

One type of x-ray source, characterized by Sco X-1 and Cyg X-2, exhibits no detectable radio noise; but a second type, characterized by Tau A, is a strong radio emitter. In the cases of Cas A, Cas B, M 87, and GX 3+1, radio sources are well centered on the apparent positions. The x-ray source GX 3+1 had been located within a 17 minute error circle including the radio source (Goss and Shaver, 1968), and the latter contained an ultraviolet object near its center as well as a faint nebula nearby: a set of unusual features characteristic of the association of x-ray sources with supernovae.

Nevertheless, identifications are very difficult. Most of the x-ray sources are not associated with radio sources, and can only be expected to appear as extremely faint optical sources, of which the sky has so many that accidental coincidences are too likely. A clear necessity is the obtaining of very accurate angular coordinates for the still unidentified x-ray objects. This can best be done with image-forming telescopes based on total reflection of soft x-rays by focussing mirrors (Giacconi, Gursky, and van Speybroeck, 1968).

7.4. Distribution in Space

Almost all the x-ray sources have angular coordinates within 3° of the plane of the Galaxy. Clearly, therefore, (a) they are galactic objects and (b) their average distance from the Sun is large compared with the thickness of the disk. There are only four sources found thus far with galactic declination more than 10°. One of these cases is Sco X-1, for which the high latitude (24°) is accounted for by proximity: this source produces the most intense flux at the Earth. Another is M87, an exception to the rule: the identification implies that the source is a remote galaxy. The other two cases (Leo X-1 and Lyr X-1) are not yet explained.

Within the disk, the known sources are not distributed uniformly, but form two concentrations. One of these is in the Cygnus–Cassiopeia region, galactic longitude $l_{\mathrm{II}} = 60°–120°$, associated with the Orion–Cygnus arm, which includes the Sun. The other region, $l_{\mathrm{II}} = 315°$ to 40°, is associated with the Sagittarius arm, the neighboring spiral arm in the direction inwards towards the galactic center. Thus, we are not seeing sources as far away as the center of the Galaxy, but only those that are close enough to produce a flux above the limits of detectability. Friedman, Byram, and Chubb (1967) estimate the average distances to be 1250 pc to the sources in the Orion arm, and 2500 pc to the sources in Sagittarius, with 150 pc the mean distance of the sources from the galactic plane. Friedman *et al.* then compute the mean luminosity of the sources in the 1–10 Å range to be 6×10^{36} ergs/sec, and the number of similar sources in the Galaxy to be 1250; thus the total luminosity of the Galaxy in this wavelength interval is found to be about 7×10^{39} ergs/sec.

A comparison with high-energy gamma radiation is of some interest. If one takes the gamma-ray flux associated with the galactic disk, as recorded by Clark *et al.* on OSO-III, and integrates it over all angles, one can find the average gamma-ray density in the disk, and from this the strength of the Galaxy as a gamma-ray source. Assuming the mean energy of the gammas above 100 MeV to be 300 MeV, I compute that the Galaxy emits 2×10^{39} ergs per second in gamma rays above 100 MeV. [The relation is $Q = (c\tau)^{-1} V \bar{E} \int J \, d\Omega$, where V is the volume of the disk and $c\tau$ the average path length from a point in the disk to the surface.] Despite the approximate nature of these estimates and the steepness of the spectra of *some* of the sources within the Galaxy, one sees that on the whole, the radiation spectrum in the Universe does not fall off rapidly with energy, but is quite hard.

A view of a galaxy from the outside was provided recently by the detection of x-rays from the Large Magellanic Cloud (Mark *et al.*, 1969).

The individual sources within the galaxy were too faint to observe at that distance (about 50 kpc), but the galaxy as a whole appeared as a faint source of 12° angular diameter. The total x-ray emission between 1.5 and 10.5 keV was evaluated at 4×10^{38} ergs/sec, about 6% of Friedman's estimate of the luminosity of our Galaxy. Since the mass of the LMC is an order of magnitude less than the mass of our Galaxy, the relative populations of x-ray sources in these two galaxies appear to be similar.

7.5.   Dimensions of the Sources

Direct information about the size of the x-ray emitting object is only available for two sources, Scorpius X-1 and the Crab Nebula. As mentioned above, the x-ray source in Scorpius was shown to have an angular diameter of less than 20 arc seconds, or 10^{-4} radians (Gursky et al., 1966b). Since the distance is about 10^{21} cm, this only implies that the diameter is less than 10^{17} cm. The point-like image of the optical counterpart implies a diameter at least 20 times smaller than this. But the sharpest indication of size comes from the observation (Lewin et al., 1968) of a major change of intensity—a factor of four—occurring in ten minutes. This implies an upper limit of a few times 10^{13} cm for the dimensions.

The Crab Nebula, on the contrary, exhibits an emitting region of about 100 arc-second diameter, implying a linear diameter of almost three light-years. This is consistent with the absence of measurable variations in x-ray luminosity during the few years in which the source has been observed.

However, the above statement is only partially true. About ten percent of the flux is modulated periodically, showing pulse rise times on the order of a millisecond and a period of 30 msec. It is virtually certain that the periodicity is due to rotation, and an object that rotates 30 times per second cannot have a radius more than 1.6×10^8 cm without exceeding the speed of light.

Thus, x-ray sources vary in size from very small to very large. The dimensions of the source in M87 are probably on the order of 100×1000 pc, the size of the optical jet.

7.6.   Spectra

Much effort has been applied to measurement of the x-ray spectra as a clue to the nature of the sources. There are three characteristic forms of continuous x-ray spectrum that would be indicative of possible conditions:

(a) Black body: a thermal radiator that is thick to its own radiation. The number of photons per unit energy interval varies as

$$dJ(\nu) \approx \nu^2 e^{-h\nu/kT}\,d\nu.$$

(b) Thin, hot plasma: a thermal radiator of low density, transparent to the radiation.

$$dJ(\nu) \approx \nu^{-1} e^{-h\nu/kT}\,d\nu.$$

(c) Synchrotron source: If the radiation is magnetic bremsstrahlung, the spectrum can have a great variety of shapes, including exponential ones like those above, in cases where the electron spectrum is cut off at high energy. But there is a tendency in nature for power-law spectra to rise, having the form

$$dJ(\nu) \approx \nu^{-(\alpha+1)}\,d\nu$$

with α a slowly-varying number, related to the exponent γ in the electron spectrum by $\gamma = 2\alpha + 1$.

Of course, a structured thermal radiator containing regions of different temperature may differ in spectral shape from the forms (a) and (b), even having the possibility of simulating a synchrotron power-law spectrum. Shklovsky's (1967) model of the Sco X-1 source postulates multiple layers at different temperature, while Sartori and Morrison (1967) even account for the apparent power-law spectrum of the Crab Nebula in terms of thermal radiation.

Furthermore, the passage of the radiation through space may modify the spectrum by absorption; and even within the source, attenuation may become important at long wavelengths. The absorption coefficient is a rapidly varying function of the quantum energy, going as $E^{-8/3}$ between absorption edges, where there are sharp changes associated with ionization potentials of particular elements. Thus, it is possible to gain information about the distances of the sources and the density and composition of the interstellar gas if one can recognize distortion attributable to absorption in the spectra of various sources.

The anticipated absorption in interstellar space has been calculated by Felten and Gould (1966), and more recently by Bell and Kingston (1967). For a source at a distance of a few kiloparsec the absorption is expected to be serious at energies below about 1.5 kilovolts. At a wavelength of 50 Å, one expects absorption effects even at distances of 100 to 200 parsecs: i.e., even for the nearest extrasolar x-ray source.

The present status of the experimental information on spectra is complicated. The data are much too voluminous to present here, or even to interpret with any degree of completeness. In general, the sources are of two types. One type, characterized by the Crab Nebula, Cygnus X-1, and M87, have spectra that fit a power law over a wide range of energies. The other type, characterized by Sco X-1 and Cyg X-2, have spectra of exponential shape that could be consistent with either a black body or a thin plasma, though the latter is considered more probable on other grounds. The temperature that fits the Scorpius source is $(5 \text{ to } 6) \times 10^7$ °K ($kT \cong 5$ keV), while that which fits the Centaurus source is three times lower.

However, when the spectra are examined in detail they are not so simple. For instance, the thermal spectra are not a really good fit to the form predicted for single temperatures. Nor would it be reasonable to expect such big and tumultuous structures to be in thermal equilibrium. Hard radiation that could not be produced at 6×10^7 °K has been detected from Sco X-1. The optical spectra from the visual objects associated with such sources are characterized by much lower temperatures, on the order of 10^4 degrees.

As for the absorption effects, there have been some efforts to detect the turnover of the spectrum at low quantum energies, owing to a transmission function varying as $e^{-(E_0/E)^{8/3}}$. Gorenstein $et\ al.$ (1967) have detected curvature in the spectral range 1–3 keV indicative of values of E_0 around 2 keV for two of the sources in Cygnus. Rappaport $et\ al.$ (1969) have detected absorption effects near 1 keV for sources in the Scorpius–Sagittarious cluster. But the "hard" sources, Cyg X-1 and Tau A, showed no absorption effects in the soft part of their spectra. Most oddly, Fritz $et\ al.$ (1968) measured x-rays of 0.25 keV from Sco X-1 without being able to detect absorption effects. Quite possibly, the interstellar medium varies in density. It is also true, however, that the emission spectra need to be more thoroughly understood before the absorption in space can be accurately assessed.

7.7.   Variability

A characteristic of x-ray sources that has appeared to be more the rule than the exception is variability. Until the present, observations have been limited to short periods of time and sources have not been kept under steady surveillance, so the form and nature of the variability is not known. What is known is that (a) the variation in intensity can be very large, (b) not only the total intensity can change, but also its spectral distribution, and (c) the changes include both long-term effects and short, flare-like variations.

When Cen X-2 was first observed (Harries *et al.* 1967), its flux at long wavelengths (5–10 Å) exceeded even that of Sco X-1 and its temperature was about 4×10^7 °K. Within a month the apparent temperature had dropped a factor of two and the flux had diminished more than a factor of three. Values of kT characterizing the soft spectrum of Sco X-1 have been reported as varying between 4 keV and 7 keV. In the flare mentioned above (Lewin *et al.*, 1968), the change of flux in at least part of the x-ray spectrum was a factor of five, but seemed to be smaller in other parts of the spectrum. The optical object associated with Sco X-1 is a variable, but exhibited no correspondingly large flare at the same time. Other sources that have been observed to show strong variations include Cyg X-2 and Cyg X-1 (Overbeck and Tananbaum, 1968).

7.8.   Two Types of Source

On the bases of dimensions, spectral shape, variability, and association with radio emission, the x-ray sources fall into two strongly contrasting groups; although, as is usual in astronomy, one needs a third class for objects that are not clearly of one type or the other.

(1) Extended sources

These are characterized by Tau A, Cas A, Cas B, M87, and the background radiation. To the extent known, these have power-law x-ray spectra, are associated with strong radio sources, and have long time constants for change. The emission mechanism may be synchrotron radiation.

(2) Compact sources

These are characterized by Sco X-1, Cyg X-2, and Cen X-2. They show no evidence of radio emission. They exhibit exponential spectra corresponding to the bremsstrahlung emission process at high temperatures, 10^7–10^8 °K. These temperatures must exist in a gaseous envelope: the associated stellar object not only is weak in the visible, but has atomic excitation lines corresponding to $T \cong 10^4$ °K. The sources are rapid variables in both visible and x-ray parts of the spectrum, on time scales of months, days, hours and even minutes.

(3) Exceptions

Lup X-1 and Cyg X-1 have hard spectra, more of the power-law type than exponential; yet they are not associated with radio-emitting objects; and Cyg X-1 has exhibited time variations on a short time scale. On the other hand, the Lupus source has been under observation for forty days from a satellite with no indication of fluctuations in intensity. There are probably many sources that do not submit to simple characterization.

They may have a multilayer structure with different temperatures, and their spectra may be strongly influenced by internal absorption.

X-ray emission by the extended sources is not simply or directly related to the radio emission. For instance, Cygnus A, one of the strongest radio sources, has not been successfully detected in the x-ray band. The galactic center (Sgr A) is also a weak x-ray source. It may be true that all supernovae are x-ray emitters, but the radio flux from Cas A exceeds that from the Crab Nebula, while the x-rays are at least ten times less intense than those from the Crab. This difference is attributable to slope of the electron spectrum and to distance: Cas A is three times farther away. The failure to detect x-rays from the Kepler supernova (1604) may simply be due to its being about ten times farther away than the Crab Nebula.

7.9.   Binary Nature of Sources

The optical counterparts of the compact x-ray sources resemble old novae, which are believed to be binary in nature.

Spectral data on the Cyg X-2 source show particularly strong evidence for a rotating system. Burbidge, Lynds, and Stockton (1967) derived the following velocities, relative to the Sun, of the matter responsible for absorption lines (H and CaII) and for an emission line (HeII 4686) at different times.

	Apr. 9	May 7	June 9	July 2	
Absorption	+ 75	+152	−387	− 96	km/sec
Emission	−329	−105	− 89	−410	

These data indicate a rapidly rotating system with $v \cong 250$ km/sec. However, the spectral data are very complex and do not fit a simple binary system (see e.g., Kraft and Demoulin, 1967, for further elaboration). The velocities are not sinusoidal functions of time, and on any one date there is a wide spread in velocity for the absorption lines. It has been suggested that the system has more than two components. The spectral complexity may be associated with emission and absorption in rapidly moving gas streams in a close binary system.

Old binaries frequetly have a white-dwarf component, and it may be that the strongest sources have a component that is even smaller, namely a neutron star, as suggested by Shklovsky (1967). If one component is very compact, the loss of gravitational energy as gas falls towards the smaller component provides sufficient energy for maintaining the temperatures needed to account for the x-ray emission.

However, S. Hayakawa and D. Sugimoto (1967) have developed a model appropriate to white-dwarf binaries, according to which the accretion of matter acts as a trigger rather than the sole source of thermal energy. A nova with a degenerate core of heavy elements still has unconsumed hydrogen near the surface. Heating of this envelope by gravitational accretion can initiate instable nuclear burning of the hydrogen ("hydrogen flicker"), releasing a total energy thousands of times greater than the gravitational energy of the infalling gas. The thermal instability in the envelope produces a shock wave that transfers energy to the corona, producing high-energy particles. Stability is then restored at the surface until further accretion triggers another nova-like flare. A model of this sort not only provides a generous supply of energy, capable of producing coronal temperatures of 10^8 °K, but also accounts for variability and flare-like behavior of x-ray sources, and for optical features such as the blue shift of absorption lines. Since this model does not require a neutron star, it also helps to explain the large number of x-ray sources present in the Galaxy.

7.10. Neutron Stars and Pulsars

Subsequent to the presentation of these lectures came the discovery of short-period pulsars in two supernovae, the Crab Nebula and Vela X (Staelin and Reifenstein, 1968; Large, Vaughan, and Mills, 1968), with periods of 33 and 89 milliseconds, respectively. This was followed by the discovery of optical pulsing in the Crab Nebula (Cocke, Disney, and Taylor, 1969), and finally of x-ray pulsations synchronized with the optical ones (Fritz, *et al.* 1969; Bradt *et al.*, 1969). Also it was found that the pulsars are slowing down.

The stability of the pulsar periods makes it clear that the periodicity is based on rotation, and the smallness of the period implies that the object is a neutron star rather than a white dwarf. When the star contracted, loss of gravitational energy produced the energy of the rapid rotation, while the mechanical rotation provides dissipative mechanisms that can explain the slowing down. The rate of loss of rotational energy is sufficient to maintain the total radiant output of the Crab Nebula.

The condensation of matter certainly produced an extremely strong magnetic field around the neutron star, by means of which the star compels the sorrounding plasma to rotate with it. At a radius not much more than 10^8 cm, the rotation becomes relativistic. Either by this direct means or by radiation fields from a rotating dipole that is not parallel to the axis of rotation, particles are accelerated to relativistic velocity. Thus

8*

the energy transference begins by production of relativistic particles, which in turn can heat and cause turbulence in a wider region.

The exact mechanism by which the x-ray pulsations are produced is unknown. But the above discoveries have given the most definite picture of the process of generation of fast charged particles that the world has seen. The phenomenon is not unique: very many stars must go through the process of collapse to the state of neutron matter, while others condense to white dwarfs and still others condense unstably beyond the Schwarzschild limit. There now exists a new starting point for thinking about the origin of cosmic-ray particles, gamma rays, and x-rays. The implications are rapidly unfolding, but were not available in time for inclusion in the lectures or in these long-delayed notes.

REFERENCES

ABRAHAM, F., J. KIDD, M. KOSHIBA, R. LEVI SETTI, C. H. TSAO, and W. WOLTER (1963), Nuovo Cimento **28**, 221.

ABRAHAM, P. B., K. A. BRUNSTEIN, and T. L. CLINE (1966), Phys. Rev. **150**, 1088.

ANAND, K. C., R. R. DANIEL, and S. A. STEPHENS (1968*a*), Phys. Rev. Letters **20**, 764.

ANAND, K. C., R. R. DANIEL, and S. A. STEPHENS (1968*b*), Proc. Ind. Acad. Sci. **48**A, 219.

ANAND, K. C., R. R. DANIEL, and S. A. STEPHENS (1968*c*), Proc. Ind. Acad. Sci. **47**A, 267.

ANAND, K. C., R. R. DANIEL, and S. A. STEPHENS (1968*d*), Can. J. Phys. **46**, S 484.

ANDERSON, K. A. (1965), *Proceedings of the Ninth International Conference on Cosmic Rays, London, 1965* (The Institute of Physics and the Physical Society, London), p. 250.

ANDREW, B. H., C. R. PURTON, and Y. TERZIAN (1967), Nature **215**, 493.

APPARAO, M. V. K. (1967), Proc. Ind. Acad. Sci. **65**A, 349.

ARNOLD, J. R., A. E. METZGER, E. C. ANDERSON, and M. A. VAN DILLA (1962), J. Geophys. Res. **67**, 4878.

BAXTER, A. J., B. G. WILSON, and D. W. GREEN (1969), Can. J. Phys. **47**, 2651.

BEEDLE, R., and W. R. WEBBER (1968), Can. J. Phys. **46** S 1014.

BELL, K. L., and A. E. KINGSTON (1967), Monthly Notices Roy. Astron. Soc. **136**, 241.

BERGAMINI, R., P. LONDRILLO, and G. SETTI (1968), Nuovo Cimento **53**B, 518.

BEUERMANN, K. P., C. J. RICE, E. C. STONE, and R. E. VOGT (1969), Phys. Rev. Letters **22**, 412.

BLANCO, V., W. KUNKEL, W. A. HILTNER, G. LYNGA, H. BRADT, G. CLARK, S. NARANAN, S. RAPPAPORT, and G. SPADA (1968), Astrophys. J. **152**, 1015.

BLAND, C. J., G. BOELLA, G. DEGLI ANTONI, C. DILWORTH, L. SCARSI, G. SIRONI, B. AGRINIER, Y. KOECHLIN, B. PARLIER, and J. VASSEUR (1966), Phys. Rev. Letters **17**, 813.

BLEEKER, J. A. M., J. J. BURGER, A. J. M. DEERENBERG, A. SCHEEPMAKER, B. N. SWANENBURG, Y. TANAKA, S. HAYAKAWA, F. MAKINO, and H. OGAWA (1968), Can. J. Phys. **46**, S 461.

BLEEKER, J. A. M., J. J. BURGER, A. J. M. DEERENBERG, A. SCHEEPMAKER, B. N. SWANENBURG, and Y. TANAKA (1968), Can. J. Phys. **46**, S 522.

BORTOLOT, V. J., JR., J. F. CLAUSER, and P. THADDEUS (1969), Phys. Rev. Letters 22, 307.

BOWYER, S., E. T. BYRAM, T. A. CHUBB, and H. FRIEDMAN (1964), Science 146, 912.

BOWYER, C. S., G. B. FIELD, and J. E. MARK (1968), Nature 217, 32.

BRADT, H., W. MAYER, S. NARANAN, S. RAPPAPORT, and G. SPADA (1967), Astrophys. J. 150, L199.

BRADT, H., S. NARANAN, S. RAPPAPORT, and G. SPADA (1968), Astrophys. J. 152, 1005.

BRADT, H., S. RAPPAPORT, W. MAYER, R. E. NATHER, B. WARNER, M. MAC FARLANE. and J. KRISTIAN (1969), Nature 222, 728.

BRECHER, K., and P. MORRISON (1967), Astrophys. J. 150, L61.

BRUNSTEIN, K. A. (1965), Phys. Rev. 137, 759.

BRYANT, D. A., T. L. CLINE, U. D. DESAI, and F. B. McDONALD (1962), J. Geophys. Res. 67, 4983.

BURBIDGE, E. M., C. R. LYNDS, and A. N. STOCKTON (1967), Astrophys. J. Letters 150, L95.

BYRAM, E. T., T. A. CHUBB, and H. FRIEDMAN (1966), Science 152, 66.

CATZ, P., J. GAWIN, J. HIBNER, F. HERBELLEAU, R. MAZE, J. WDOWCZYK, and A. ZAWADSKI (1967), Compt. Rend. 265, 84.

CHARMAN, W. N., J. V. JELLEY, P. R. ORMAN, R. W. P. DREVER, and B. McBREEN (1968), Nature 220, 565.

CHODIL, G., R. C. JOPSON, H. MARK, F. D. SEWARD, and C. D. SWIFT (1965), Phys. Rev. Letters 15, 605.

CHUDAKOV, A. E., V. L. DADYKIN, V. I. ZATSEPIN, and N. M. NESTEROVA (1962), J. Phys. Soc. Japan 17, Suppl. A III, 106; also, *Proceedings of the Fifth Interamerican Seminar on Cosmic Rays and Space Physics, La Paz, Bolivia.*

CLARK, G. W., G. P. GARMIRE, and W. L. KRAUSHAAR (1968), Astrophys. J. Letters 153, L203.

CLINE, T. L., G. H. LUDWIG, and F. B. McDONALD (1964), Phys. Rev. Letters 13, 786.

CLINE, T. L., and F. B. McDONALD (1968), Can. J. Phys. 46, S761.

CLINE, T. L., and E. W. HONES, JR. (1968), Can. J. Phys. 46, S527.

COCKE, W. J., M. J. DISNEY, and D. J. TAYLOR (1969) ,I.A.U. Circ. No. 2128.

COLGATE, S. A., and R. H. WHITE (1966), Astrophys. J. 143, 626.

COWSIK, R., Y. PAL, S. N. TANDON, and R. P. VERMA (1966), Phys. Rev. Letters 17, 1298.

COWSIK, R., and Y. PAL (1969), Phys. Rev. Letters 22, 550.

CRITCHFIELD, C. L., E. P. NEY, and S. OLEKSA (1952), Phys. Rev. 85, 461.

DANIEL, R. R., and S. A. STEPHENS (1966), Phys. Rev. Letters 17, 935.

DANJO, A., S. HAYAKAWA, F. MAKINO, and Y. TANAKA (1968), Can. J. Phys. 46, S530.

DELVAILLE, J. P., P. ALBATS, K. I. GREISEN, and H. B. ÖGELMAN (1968), Can. J. Phys. 46, S425.

DESHONG, J. A., JR., R. H. HILDEBRAND, and P. MEYER (1964), Phys. Rev. Letters 12, 3.

DUTHIE, J. G. (1968), Can. J. Phys. 46, S401.

DUTHIE, J. G., P. H. FOWLER, A. KADDOURA, D. H. PERKINS, and K. PINKAU (1962), Nuovo Cimento 24, 122.

EARL, J. A. (1961), Phys. Rev. Letters 6, 125.

EGGEN, O. J., K. C. FREEMAN, and A. SANDAGE (1968), Astrophys. J. Letters 154, L27.

ENCRENAZ, P., and R. B. PARTRIDGE (1969), Astrophys. Letters 3, 161.

FAN, C. Y., G. GLOECKLER, and J. A. SIMPSON (1964), Phys. Rev. Letters 13, 149.

FAN, C. Y., G. GLOECKLER, and J. A. SIMPSON (1965), *Proceedings of the Ninth International Conference on Cosmic Rays, London, 1965*, p. 105.

FAN, C. Y., G. GLOECKLER, J. A. SIMPSON, and S. D. VERMA (1968), Astrophys. J. **151**, 737.

FANSELOW, J. L. (1968), Astrophys. J. **152**, 783.

FAZIO, G. G. (1967), Ann. Rev. Astron. Astrophys. **5**, 481.

FAZIO, G. G., and H. F. HELMKEN (1968), Can. J. Phys. **46**, S456.

FAZIO, G. G., H. F. HELMKEN, S. J. CAVRAK, JR., and D. R. HEARN (1968), Can. J. Phys. **46**, S427.

FAZIO, G. G., H. F. HELMKEN, G. H. RIEKE, and T. C. WEEKES (1968a), Astrophys. J. Letters **154**, L83.

FAZIO, G. G., H. F. HELMKEN, G. H. RIEKE, and T. C. WEEKES (1968b), Can. J. Phys. **46**, S451.

FAZIO, G. G., H. F. HELMKEN, G. H. RIEKE, and T. C. WEEKES (1968c), Nature **220**, 892.

FAZIO, G. G., H. F. HELMKEN, G. H. RIEKE, and T. C. WEEKES (1969), I.A.U. Symposium No. 37 on Non-Solar X and Gamma-Ray Astronomy, Rome, Italy.

FEGAN, D. J., B. MCBREEN, E. P. O'MONGAIN, N. A. PORTER, and P. J. SLEVIN (1968), Can. J. Phys. **46**, S433.

FELTEN, J. E. (1966), Astrophys. J. **145**, 589.

FELTEN, J. E., and P. MORRISON (1963), Phys. Rev. Letters **10**, 453.

FELTEN, J. E., and P. MORRISON (1966), Astrophys. J. **146**, 686.

FELTEN, J. E., and R. J. GOULD (1966), Phys. Rev. Letters **17**, 401.

FELTEN, J. E. (1968), Astrophys. J. **151**, 861.

FERMI, E. (1949), Phys. Rev. **75**, 1169.

FICHTEL, C. E., and F. B. MCDONALD (1967), Ann. Rev. Astron. Astrophys. **5**, 351–398.

FICHTEL, C. E. (1968), Sky and Telescope **35**, No. 2, p. 1.

FICHTEL, C. E., T. L. CLINE, C. H. EHRMANN, D. A. KNIFFEN, and R. W. ROSS (1968), Can. J. Phys. **46**, S419.

FICHTEL, C. E., and D. A. KNIFFEN (1965), J. Geophys. Res. **70**, 4227.

FICHTEL, C. E., D. A. KNIFFEN, and H. B. ÖGELMAN (1969), Astrophys. J. **158**, 193.

FISHMAN, G. J., F. R. HARNDEN, JR., and R. C. HAYMES (1969), Astrophys. J. Letters **156**, L107.

FRANK, L. A., J. A. VAN ALLEN, and E. MACAGNO (1963), J. Geophys. Res. **68**, 3543.

FRIEDMAN, H., and E. T. BYRAM (1967), Science **158**, 257.

FRIEDMAN, H., E. T. BYRAM, and T. A. CHUBB (1967), Science **156**, 374.

FRITZ, G., J. F. MEEKINS, R. C. HENRY, E. T. BYRAM, and H. FRIEDMAN (1968) Astrophys. J. Letters **153**, L199.

FRITZ, G., R. C. HENRY, J. F. MEEKINS, T. A. CHUBB, and H. FRIEDMAN (1969), Science **164**, 709.

FRUIN, J. H., J. V. JELLEY, C. D. LONG, N. A. PORTER, and T. C. WEEKES (1964), Phys. Letters **10**, 176.

FRYE, G. M., Jr., and L. H. SMITH (1966), Phys. Rev. Letters **17**, 733.

FRYE, G. M., Jr., and C. P. WANG (1968), Can. J. Phys. **46**, S 448.

FRYE, G. M., Jr., and C. P. WANG (1969), Astrophys. J. **158**, 925.

GARMIRE, G., and W. KRAUSHAAR (1965), Space Sci. Rev. **4**, 123.

GAWIN, J., J. HIBNER, R. MAZE, J. WDOWCZYK, and A. ZAWADSKI (1965), *Proceedings of the International Conference on Cosmic Rays, London* (The Institute of Physics and the Physical Society, London), Vol. 2, p. 639.

GIACCONI, R., H. GURSKY, and L. P. VAN SPEYBROECK (1968), Ann. Rev. Astron. Astrophys. **6**, 373.

GIACCONI, R., H. GURSKY, F. PAOLINI, and B. ROSSI (1962), Phys. Rev. Letters 9, 439.

GIACCONI, R., P. GORENSTEIN, H. GURSKY, P. D. USHER, J. R. WATERS, A. SANDAGE, P. OSMER, and J. V. PEACH (1967a), Astrophys. J. Letters 148, L129.

GIACCONI, R., P. GORENSTEIN, H. GURSKY, and J. R. WATERS (1967b), Astrophys. J. Letters 148, L119.

GINZBURG, V. L., and S. I. SYROVATSKII (1964), The Origin of Cosmic Rays (Pergamon Press, Macmillan Co., New York).

GORENSTEIN, P., R. GIACCONI, and H. GURSKY (1967), Astrophys. J. Letters 150, L85.

GORENSTEIN, P., E. M. KELLOGG, and H. GURSKY (1969), Astrophys. J. 156, 315.

GOSS, W. M., and P. A. SHAVER (1968), Astrophys. J. Letters 154, L75.

GOULD, R. J. (1965), Phys. Rev. Letters 15, 577.

GOULD, R. J. (1967), Am. J. Phys. 35, 376.

GOULD, R. J., and G. R. BURBIDGE (1965), Ann. Astrophys. 28, 171.

GOULD, R. J., and G. R. BURBIDGE (1967), in Handbuch der Physik, edited by S. Flugge (Springer-Verlag, Berlin), Vol. XLVI/2, p. 265.

GOULD, R. J., and G. P. SCHRÉDER (1966), Phys. Rev. Letters 16, 252.

GOULD, R. J., and G. P. SCHRÉDER (1967), Phys. Rev. 155, 1408.

GREISEN, K. (1966a), in Perspectives in Modern Physics (Interscience Publishers, New York), p. 355.

GREISEN, K. (1966b), Phys. Rev. Letters 16, 748.

GREISEN, K. (1967), in Topics in Relativistic Astrophysics, Proceedings of the 1967 Texas Symposium, edited by A. G. W. Cameron and S. P. Maran (Gordon and Breach, New York, 1970).

GURSKY, H., R. GIACCONI, P. GORENSTEIN, J. R. WATERS, M. ODA, H. BRADT, G. GARMIRE, and B. V. SREEKANTAN (1966a), Astrophys. J. 146, 310.

GURSKY, H., R. GIACCONI, P. GORENSTEIN, J. R. WATERS, M. ODA, H. BRADT, G. GARMIRE, and B. V. SREEKANTAN (1966b), Astrophys. J. 144, 1249.

HARRIES, J. R., K. G. McCRACKEN, R. J. FRANCEY, and A. G. FENTON (1967), Nature 215, 38.

HARTMAN, R. C., R. H. HILDEBRAND, and P. MEYER (1965), J. Geophys. Res. 70, 2713.

HARTMAN, R. C. (1967), Astrophys. J. 150, 371.

HASEGAWA, H., K. MURAKAMI, S. SHIBATA, K. SUGA, Y. TOYODA, V. DOMINGO, I. ESCOBAR, K. KAMATA, H. BRADT, G. CLARK, and M. LA POINTE (1965), Proceedings of the International Conference on Cosmic Rays, London (The Institute of Physics and the Physical Society, London), p. 708.

HAYAKAWA, S., and M. MATSUOKA (1964), Progr. Theoret. Phys. (Kyoto) Suppl. 30, 204.

HAYAKAWA, S., and D. SUGIMOTO (1967), Preprint, Nagoya University.

HAYMES, R. C., and W. CRADDOCK (1966), J. Geophys. Res. 71, 3261.

HAYMES, R. C., D. V. ELLIS, G. J. FISHMAN, J. D. KURFESS, and W. H. TUCKER (1968a), Astrophys. J. Letters 151, L9.

HAYMES, R. C., D. V. ELLIS, G. J. FISHMAN, S. W. GLENN, and J. D. KURFESS (1968b), Astrophys. J. Letters 151, L125.

HAYMES, R. C., D. V. ELLIS, G. J. FISHMAN, S. W. GLENN, and J. D. KURFESS (1968c), Astrophys. J. Letters 151, L131.

HENRY, R. C., G. FRITZ, J. F. MEEKINS, H. FRIEDMAN, and E. T. BYRAM (1968), Astrophys. J. Letters 153, L11.

HILLAS, A. M. (1968), Can. J. Phys. 46, S623.

HOUCK, J. R., and M. HARWIT (1969), Astrophys. J. **157**, L45.

HULSIZER, R. I., and B. ROSSI (1948), Phys. Rev. **73**, 1402.

ISRAEL, M. H. (1969), J. Geophys. Res. **74**, 4701.

JACOBSON, A. S. (1968), Thesis, University of California at San Diego.

JELLEY, J. V. (1966), Phys. Rev. Letters **16**, 479.

JOKIPII, J. R., and P. MEYER (1968), Phys. Rev. Letters **20**, 752.

JONES, F. C. (1963), J. Geophys. Res. **68**, 4399.

JONES, F. C. (1965), Phys. Rev. Letters **15**, 512.

KIDD, J. M. (1962), Nuovo Cimento **27**, 57.

KIEPENHEUER, K. O. (1950), Phys. Rev. **79**, 738.

KRAFT, R. P., and M. DEMOULIN (1967), Astrophys. J. Letters **150**, L183.

KRAUSHAAR, W. L., G. W. CLARK, G. GARMIRE, H. HELMKEN, P. HIGBIE, and M. AGO-
GINO (1965), Astrophys. J. **141**, 845.

KRIMIGIS, S. M., and J. A. VAN ALLEN (1967), J. Geophys. Res. **72**, 4471.

LARGE, M. I., A. E. VAUGHAN, and B. Y. MILLS (1968), Nature **220**, 340.

LEGG, M. P. C., and K. C. WESTFOLD (1968), Astrophys. J. **154**, 499.

LEWIN, W. H. G., G. W. CLARK, and W. B. SMITH (1968), Astrophys. J. Letters **152**, L55.

L'HEUREUX, J. (1967), Astrophys. J. **148**, 399.

L'HEUREUX, J., and P. MEYER (1968), Can. J. Phys. **46**, S892.

L'HEUREUX, J., P. MEYER, S. D. VERMA, and R. VOGT (1968), Can. J. Phys. **46**, S896.

LINSLEY, J., and L. SCARSI (1962), Phys. Rev. Letters **9**, 123.

LINSLEY, J. (1962), Phys. Rev. Letters **9**, 126.

LONGAIR, M. S. (1966), Monthly Notices Roy. Astron. Soc. **133**, 421.

LÜST, R., and K. PINKAU (1968), in *Electromagnetic Radiation in Space*, edited by J. G.
Emming (D. Reidel Publishing Co., Dordrecht, Holland).

MARASCHI, L., G. C. PEROLA, and S. SCHWARZ (1968), preprint.

MARK, H., R. PRICE, R. RODRIGUES, F. D. SEWARD, and C. D. SWIFT (1969), Astrophys.
J. Letters **155**, L143.

MAY, T. C. (1968), Ph. D. Thesis, University of Minnesota; available as Technical Report
CR-118.

MAY, T. C., and C. J. WADDINGTON (1969), Astrophys. J. **156**, 437.

METZGER, A. E., E. C. ANDERSON, M. A. VAN DILLA, and J. R. ARNOLD (1964), Nature
204, 766.

MEYER, P., and R. VOGT (1961), Phys. Rev. Letters **6**, 193.

MORRISON, P. (1958), Nuovo Cimento **7**, 858.

MORRISON, P. (1961), "The Origin of Cosmic Rays", in *Handbuch der Physik*, Cosmic
Rays I, edited by S. Flugge (Springer-Verlag, Berlin), Vol. XLVI/1.

MURAYAMA, T., and J. A. SIMPSON (1968), J. Geophys. Res. **73**, 891.

NIKISHOV, A. I. (1961), Zh. Eksp. Teor. Fiz. **41**, 549 [Sov. Phys.—JETP **14**, 393].

O'CONNELL, R. F. (1966), Phys. Rev. Letters **17**, 1232.

O'CONNELL, R. F., and S. D. VERMA (1969), Phys. Rev. Letters **22**, 1443.

ODA, M., H. BRADT, G. GARMIRE, G. SPADA, B. V. SREEKANTAN, H. GURSKY, R. GIAC-
CONI, P. GORENSTEIN, and J. R. WATERS (1967), Astrophys. J. Letters **148**, L5.

ÖGELMAN, H. (1969), Nature **221**, 753.

ÖGELMAN, H. B., J. P. DELVAILLE, and K. I. GREISEN (1966), Phys. Rev. Letters **16**, 491.

O'MONGAIN, E. P., N. A. PORTER, J. WHITE, D. J. FEGAN, D. M. JENNINGS, and B. G.
LAWLESS (1968), Nature **219**, 1348.

OVERBECK, J. W., and H. D. TANANBAUM (1968), Astrophys. J. **153**, 899.

PAL, Y. (1966), "Cosmic Rays and Their Interactions", in *Handbook of Physics*, edited by Condon and Odishaw (McGraw-Hill, New York), Second Edition.

PARKER, E. N. (1958), Phys. Rev. **110**, 445.

PARKER, E. N. (1965), Planetary and Space Sci. **13**, 9.

PEROLA, G. C., and L. SCARSI (1966), Nuovo Cimento **46**, 718.

PEROLA, G. C., L. SCARSI, and G. SIRONI (1967), Nuovo Cimento **52**B, 455.

PEROLA, G. C., L. SCARSI, and G. SIRONI (1968), Nuovo Cimento **53**B, 459.

PETERSON, L. E., A. S. JACOBSON, R. M. PELLING, and D. A. SCHWARTZ (1968), Can. J. Phys. **46**, S437.

PINE, J., R. J. DAVISSON, and K. GREISEN (1959), Nuovo Cimento **14**, 1181.

POLLACK, J. B., and G. G. FAZIO (1963), Phys. Rev. **131**, 2684.

Proceedings of the Tenth International Conference on Cosmic Rays, Calgary, Alberta, 1967; Can. J. Phys. **46**, No. 10, Parts 2–4 (1968).

Progress of Theoretical Physics, Suppl. No. 30 (1964): *Origin of Cosmic Rays*.

RAMATY, R., and R. E. LINGENFELTER (1966a), Phys. Rev. Letters **17**, 1230.

RAMATY, R., and R. E. LINGENFELTER (1966b), J. Geophys. Res. **71**, 3687.

RAMATY, R., and R. E. LINGENFELTER (1968), Phys. Rev. Letters **20**, 120.

RAPPAPORT, S., H. V. BRADT, and W. MAYER (1969), Astrophys. J. Letters **157**, L21.

RICHARDS, D., and J. M. COMELLA (1969), Nature **222**, 551.

RIEKE, G. H., and T. C. WEEKES (1969), Astrophys. J. **155**, 429.

ROCKSTROH, J., and W. R. WEBBER (1969), J. Geophys. Res. **74**, 5041.

ROSSI, B. B. (1968), International Symposium on Contemporary Physics, Trieste, Italy.

RUBTSOV, V. I., and V. I. ZATSEPIN (1968), Can. J. Phys. **46**, S518.

SANDAGE, A. R., P. OSMER, R. GIACCONI, P. GORENSTEIN, H. GURSKY, J. WATERS, H. BRADT, G. GARMIRE, B. V. SREEKANTAN, M. ODA, K. OSAWA, and J. JUGAKU (1966), Astrophys. J. **146**, 316.

SARTORI, L., and P. MORRISON (1967), Astrophys. J. **150**, 385.

SCHEIN, M., W. P. JESSE, and E. O. WOLLAN (1941), Phys. Rev. **59**, 615.

SEIELSTAD, G. A., and K. W. WEILER (1968), Astrophys. J. **154**, 817.

SEWARD, F., G. CHODIL, H. MARK, C. SWIFT, and A. TOOR (1967), Astrophys. J. **150**, 845.

SHEN, C. S. (1967), Phys. Rev. Letters **19**, 399.

SHEN, C. S. (1969), Phys. Rev. Letters **22**, 568.

SHIVANANDAN, K., J. R. HOUCK, and M. O. HARWIT (1968), Phys. Rev. Letters **21**, 1460.

SHKLOVSKY, I. S. (1967), Astrophys. J. Letters **148**, L1.

SILK, J. (1968), Astrophys. J. Letters **151**, L19.

SIMNETT, G. M., and F. B. MCDONALD (1969), Astrophys. J. **157**, 1435.

SOOD, R. K. (1969), Nature **222**, 650.

SPEISER, T. W. (1965), *Proceedings of the Ninth International Conference on Cosmic Rays, London, 1965*, p. 147.

STAELIN, D. H., and E. C. REIFENSTEIN (1968), Science **162**, 1481.

STECKER, F. W. (1968), Phys. Rev. Letters **21**, 1016.

STECKER, F. W. (1969a), Phys. Rev. **180**, 1264.

STECKER, F. W. (1969b), Goddard Space Flight Center preprint X-641-69-238.

STELJES, J. F., H. CARMICHAEL, and K. G. MCCRACKEN (1961), J. Geophys. Res. **66**, 1363.

SUGA, K., I. ESCOBAR, K. MURAKAMI, V. DOMINGO, Y. TOYODA, G. CLARK, and M. LA POINTE (1963), *Proceedings of the International Conference on Cosmic Rays, Jaipur, India*.

TUCKER, W. (1967), Astrophys. J. **148**, 745.

VALDEZ, J. V., and C. J. WADDINGTON (1969), Astrophys. J. Letters **156**, L85.

VERMA, S. D. (1967*a*), Proc. Ind. Acad. Sci. A66, 125.

VERMA, S. D. (1967*b*), Phys. Rev. Letters **18**, 253.

VERMA, S. D. (1969*a*), Can. J. Phys. **47**, 513.

VERMA, S. D. (1969*b*), Astrophys. J. Letters **156**, L79.

WALLERSTEIN, G. (1967), Astrophys. Letters **1**, 31.

WARNER, B., R. E. NATHER, and M. MACFARLANE (1969), Nature **222**, 233.

WEBBER, W. R. (1967)' "The Spectrum and Charge Composition of the Primary Cosmic Radiation", in *Handbuch der Physik, Cosmic Rays II*, edited by S. Flugge (Springer-Verlag, Berlin), Vol. XLVI/2.

WEBBER, W. R., and C. CHOTKOWSKI (1967), J. Geophys. Res. **72**, 2783.

WEBBER, W. R. (1968), Australian J. Phys. **21**, 845.

ZATSEPIN, G. T., and V. A. KUZMIN (1966), Sov. Phys.—JETP Letters **4**, 78.

Index